DeWALT®

CARPENTRY AND FRAMING

COMPLETE HANDBOOK

SECOND EDITION

GARY BRACKETT

 CENGAGE

Australia • Brazil • Mexico • Singapore • United Kingdom • United States

www.DeWALT.com/GUIDES

DEWALT Carpentry and Framing Complete Handbook, Second Edition
Gary Brackett

SVP, GM Skills & Global Product Management: Jonathan Lau

Product Director: Matthew Seeley

Senior Product Manager: Vanessa Myers

Senior Director, Development: Marah Bellegarde

Senior Product Development Manager: Larry Main

Content Developer: Jenn Alverson

Vice President, Marketing Services: Jennifer Ann Baker

Marketing Manager: Scott Chrysler

Senior Content Project Manager: James Zayicek

Content Project Management and Art Direction: Lumina Datamatics, Inc.

Cover image(s): antoniodiaz/Shutterstock.com; Brandon Bourdages/Shutterstock.com; brizmaker/Shutterstock.com

For product information and technology assistance, contact us at
Cengage Customer & Sales Support, 1-800-354-9706

For permission to use material from this text or product, submit all requests online at **www.cengage.com/permissions.** Further permissions questions can be e-mailed to **permissionrequest@cengage.com**

Library of Congress Control Number: 2018930178

ISBN: 978-1-337-39879-4

Cengage
20 Channel Center Street
Boston, MA 02210
USA

Cengage is a leading provider of customized learning solutions with employees residing in nearly 40 different countries and sales in more than 125 countries around the world. Find your local representative at **www.cengage.com.**

Cengage products are represented in Canada by Nelson Education, Ltd.

To learn more about Cengage, visit **www.cengage.com**

Purchase any of our products at your local college store or at our preferred online store **www.cengagebrain.com**

Notice to the Reader

Publisher and DEWALT® do not warrant or guarantee any of the products described herein or perform any independent analysis in connection with any of the product information contained herein. Publisher and DEWALT® do not assume, and expressly disclaim, any obligation to obtain and include information other than that provided to it by the manufacturer. The reader is expressly warned to consider and adopt all safety precautions that might be indicated by the activities described herein and to avoid all potential hazards. By following the instructions contained herein, the reader willingly assumes all risks in connection with such instructions. The publisher and DEWALT® make no representations or warranties of any kind, including but not limited to, the warranties of fitness for particular purpose or merchantability, nor are any such representations implied with respect to the material set forth herein, and the publisher and DEWALT® take no responsibility with respect to such material. The publisher and DEWALT® shall not be liable for any special, consequential, or exemplary damages resulting, in whole or part, from the readers' use of, or reliance upon, this material.

Printed in the United States of America

Print Number: 04   Print Year: 2021

CONTENTS

INTRODUCTION

The purpose of this book is to provide information and instruction on wood frame construction. It is not my desire to change the way the reader does his or her work unless it is currently inefficient. Hopefully the instructional text complimenting the numerous diagrams will give the reader confidence to build more-challenging projects.

In my experiences, I've noticed that carpenters are creatures of habit and are resistant to change, except when it will save them time and effort. Three carpenters can have three different opinions on the best way to approach a task, and they may all be good methods. The intent of this book is not necessarily to change one's current building habits; after all, some of the readers may be using excellent techniques already that are not mentioned in this book. The intent is to provide a good method to follow where one may not currently have an adequate understanding. Explanations in many cases are offered so the reader will understand why a particular methodology is used. Furthermore, knowing different methods of completing a task is comparable to having extra tools in a toolbox—some will work better than others in given situations.

With so many different materials, manufacturers, and techniques available, this book would be enormous in size if it took every variation into consideration. The focus, in this case, is on well-used practices of wood frame construction.

Building codes can vary regionally; therefore carpenters need to be familiar with the particular codes that have been adopted in their areas. Some of the framing illustrated within may be perfectly acceptable in most areas but unacceptable in others. For example, earthquake- and hurricane-prone areas may require extra reinforcement. Many cautions pertaining to building codes are referenced throughout this book; **know your local codes.** The code enforcement officers I have dealt with have been extremely helpful and tend to respect the person who is asking the questions. It is cheaper and easier to ask a question before starting, than to fix a code violation later on.

Influences for subject matter in this book include information I've absorbed during the past 30+ years. Sources include personal experiences as a general contractor; many books, magazines, and trade journals; internet information; and discussions with professionals including my colleagues at SUNY Delhi.

Carpenters have been around for thousands of years. Because of this, it is unlikely that any information in this book is purely original. However, my goal is to present it in an original format that is easily digested and useful.

Comments? Please contact the author brackegw@delhi.edu

Happy building...

Gary Brackett

ACKNOWLEDGMENTS

- My parents for always encouraging me no matter what
- Vinnie and Steve for giving me my first carpentry-related job
- Tim for patiently teaching me
- My colleagues at SUNY Delhi for giving me an opportunity to teach others
- Bill for his insight and suggestions
- My students for continuously challenging me and keeping me on my toes
- Floyd for steering me on the path of authorship and sharing his personal resources
- Every person from Cengage Learning who was involved with this project
- My wife and two daughters for always supporting and loving me

CHAPTER 1

HOUSES

HOUSES

In different areas of the country, house styles and building techniques are influenced by topography, climate, soil, underlying rock type, material availability, and more. Furthermore, the foundations supporting these homes vary for similar reasons. For example, climate can influence the depth to which the footing is placed, this may in turn influence whether or not a basement will be constructed, ultimately this may impact other decisions on the type of house that is built.

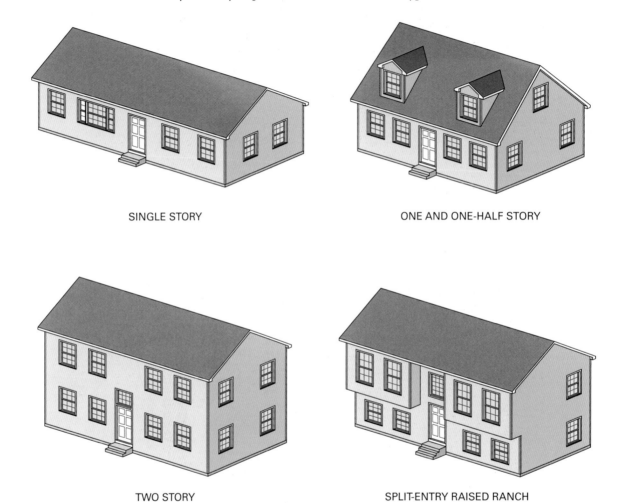

SINGLE STORY

ONE AND ONE-HALF STORY

TWO STORY

SPLIT-ENTRY RAISED RANCH

Figure 1-1 Several popular residential house styles

HOUSE TYPES

The following are a few popular house styles with some advantages/disadvantages of each.

Single Story/Ranch

Single-story houses necessitate a much larger footprint than one-and-one-half- or two-story homes; this requires more land. One advantage is that everything is on one floor, making it easier to move around, some people prefer this. Being closer to the ground, one-story houses are typically easier to construct and maintain.

Figure 1-2 Single-story house

These houses generally have shallow sloping roofs and do not have living space above the first floor.

Raised Ranch

Below is a single-story house having a basement level that is generally usable as living space. Raised ranches often have a split entry, meaning, at the main entry point, one has the choice of climbing stairs up half a story or down half a story into the living spaces. With a split entry, about half of the lower (basement) level is below grade and half above grade, the entry itself being at grade level.

Figure 1-3 Split-entry raised ranch

The advantages are: with only 4′ or so of the foundation below grade, less excavation is needed than a full basement; larger windows can easily be utilized in the basement level, allowing for more options (code-wise) for using that space. Furthermore, with a second floor of living space, the overall footprint of the house can be smaller. Often seen accompanying a raised ranch, the upper story overhangs (cantilevers) the foundation by a couple of feet, thus making the square footage of the upper story larger, without the expense of a larger foundation. This cannot be a feature of a house that has a floor level close to the ground, due to potential maintenance problems.

A significant disadvantage is that the occupants will be forced to always use the stairs upon entering the house.

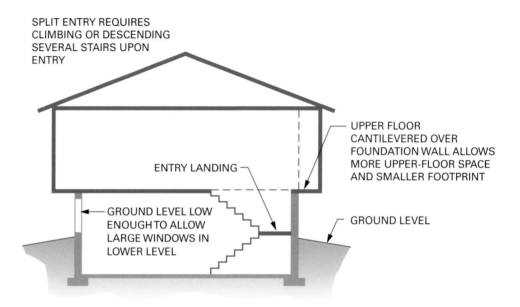

Figure 1-4 Split-entry house

One and One-Half Story

This type has a full first floor and is framed with a fairly steep roof slope, allowing for some usable floor space above the first floor. The half story, or upper story, cannot be fully utilized but has adequate headroom toward the center to use a portion of it as living space.

One-and-one-half-story houses need to have stairs and they often include dormers to allow for more natural lighting on the upper floor. By utilizing the upper half-story, the foundation footprint can be approximately 1/3 smaller than that of a single-story house, while keeping the same square footage. This may save on construction costs and/or allow the home to be built on a smaller lot.

A single-story house without a steep roof may be considered a ranch, whereas the same house with a steep sloping roof (and access to the second story) will be considered a one-and-one-half-story house. In some areas these are called Cape

Figure 1-5 One-and-one-half-story house

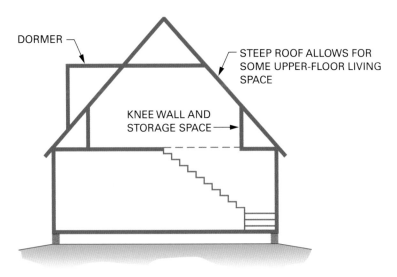

DORMER

STEEP ROOF ALLOWS FOR
SOME UPPER-FLOOR LIVING
SPACE

KNEE WALL AND
STORAGE SPACE

Figure 1-6 Cross section of a one-and-one-half-story house

Cod style homes. This is an example where simply changing the slope of the roof increases the potential usable space.

Full Two Story

These houses are more difficult to construct and maintain due to having to work from higher lifts, ladders, scaffolding, and other types of staging. Stress loads for headers and some other framing members are calculated differently than with single-story homes. Typically both stories are continuously utilized and inhabitants must be willing to climb stairs often.

Figure 1-7 Two-story house

In order to save space, stairs are generally located directly over the stairs below.

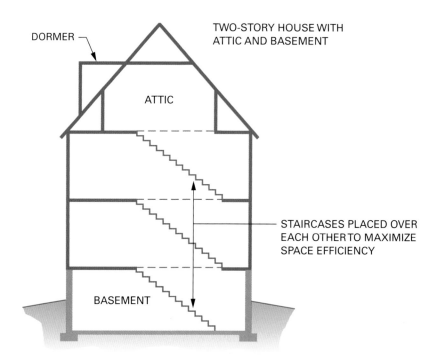

Figure 1-8 Cross section of a two-story house with attic and basement

Two-story houses can be built on approximately half the footprint of a single-story house. They will also have approximately half of the roof and foundation area. Most importantly, two-story homes can be built on a smaller piece of land than a similarly sized single-story home. This can help to save significantly on overall costs.

Other

There are other types of houses, such as salt-box houses and multi-level (split-level) houses. Advantages of these homes tend to center on topography and orientation for solar gain.

Figure 1-9 Salt-box house

SPACE AND ENERGY EFFICIENCY

Home energy use is of great importance today; therefore a few words on space and energy efficiency are necessary.

Heat Loss/Gain

Any exposed surface of a house is vulnerable to potential heat loss in cool climates, or potential heat gain in hot climates. As a result, it makes sense to consider house shapes that have less exterior surface area while simultaneously maximizing internal square footage.

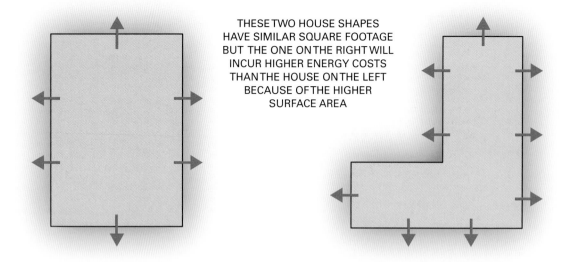

THESE TWO HOUSE SHAPES HAVE SIMILAR SQUARE FOOTAGE BUT THE ONE ON THE RIGHT WILL INCUR HIGHER ENERGY COSTS THAN THE HOUSE ON THE LEFT BECAUSE OF THE HIGHER SURFACE AREA

Figure 1-10 A house loses/gains heat from exposed surfaces

House Shape

Figures 1-11 through 1-14 represent four differently shaped structures. Through simple mathematics, it is apparent that all have the same perimeter. However, the square footage of the structures vary dramatically!

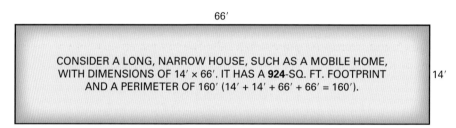

66′

CONSIDER A LONG, NARROW HOUSE, SUCH AS A MOBILE HOME, WITH DIMENSIONS OF 14′ × 66′. IT HAS A **924**-SQ. FT. FOOTPRINT AND A PERIMETER OF 160′ (14′ + 14′ + 66′ + 66′ = 160′).

14′

Figure 1-11 A mobile home is long and narrow

For rectangular objects:
Perimeter is found by measuring the distance around an object.
Area is found by multiplying the length times the width of the object.

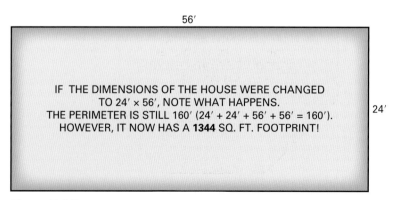

56′

IF THE DIMENSIONS OF THE HOUSE WERE CHANGED TO 24′ × 56′, NOTE WHAT HAPPENS. THE PERIMETER IS STILL 160′ (24′ + 24′ + 56′ + 56′ = 160′). HOWEVER, IT NOW HAS A **1344** SQ. FT. FOOTPRINT!

24′

Figure 1-12 Rectangular-shaped house

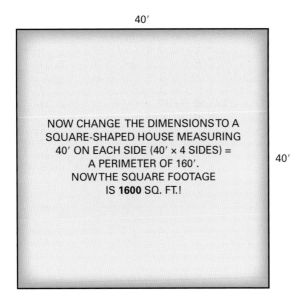

Figure 1-13 A square-shaped house

Consider Figures 1-11 through 1-14. From a material standpoint, the square footage of the floor system as well as the roof system increases in each instance. However, all three examples have the same amount of foundation and exterior wall materials. Furthermore, the heat loss/gain through the exterior walls and the labor to build the exterior walls will be similar for both the long and narrow 924 sq. ft. house and for the square-shaped 1600 sq. ft. house!

The Octagon

Octagon houses were once popular partially due to their efficient use of building materials.

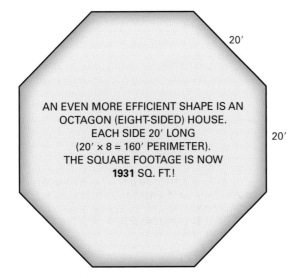

Figure 1-14 An octagon-shaped house is an efficient use of materials

Keep in mind that Figures 1-11 through 1-14 show are square feet *per floor!*

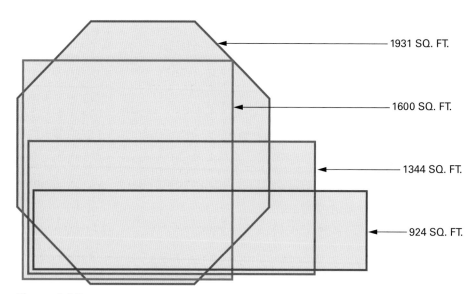

——— 1931 SQ. FT.

——— 1600 SQ. FT.

——— 1344 SQ. FT.

——— 924 SQ. FT.

Figure 1-15 These differently shaped houses have the same perimeter dimensions

The outlines in Figure 1-15 represent different house shapes that can all be built *using the same amount of exterior wall material.* As one can see in Table 1-1, some shapes are much more space/material-efficient to build than others.

The Most Efficient Shape

As it turns out, the most efficient shape is the circle. A circular house with a 160′ perimeter will have an area of **2037 sq. ft!** Building circular-shaped houses may create a host of other challenges, but when material and space efficiency are an important issue, perhaps an octagon or square shape should be considered.

EFFICIENCY AND HOUSE SHAPE

Table 1-1 Efficiency and House Shape

House Shape	Long, Narrow Mobile Home	Rectangle	Square	Octagon
Dimensions	14' × 66'	24' × 56'	40' × 40'	20' per side
Perimeter	160'	160'	160'	160'
Area (footprint)	**924** sq. ft.	**1344** sq. ft.	**1600** sq. ft.	**1931** sq. ft.
Footing	Same amount	Same amount	Same amount	Same amount
Foundation material	Same amount	Same amount	Same amount	Same amount
Framing needed for exterior walls	Same amount	Same amount	Same amount	Same amount
Sheathing needed for exterior walls	Same amount	Same amount	Same amount	Same amount
House wrap needed for exterior walls	Same amount	Same amount	Same amount	Same amount
Siding needed for exterior walls	Same amount	Same amount	Same amount	Same amount
Insulation needed for exterior walls	Same amount	Same amount	Same amount	Same amount
Labor to build exterior walls	Approximately the same	Approximately the same	Approximately the same	Approximately the same

Houses with unusual shapes, although architecturally pleasing, will be slower to build, and the added surfaces will increase the amount of materials used. This translates to an increase in exterior surface area that can add potential heat loss/gain.

A rectangular shape can be a more efficient use of materials than a house with multiple angles.

The houses represented in Figures 1-16 and 1-17 use the same amount of foundation and exterior wall materials. This will most likely make the house shown in Figure 1-17 more costly to build *per square foot* than the house represented by Figure 1-16.

As the square footage of the house increases, there will be higher costs associated with floor framing/finishing and roof framing/finishing, in both materials and labor.

The house represented in Figure 1-18 uses more material than either of the previous houses, yet is smaller in terms of square footage than the rectangular house shown in Figure 1-16. This shape, although interesting, is more wasteful in terms of material and will likely have higher energy costs associated with it.

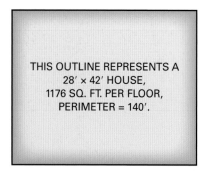

THIS OUTLINE REPRESENTS A
28′ × 42′ HOUSE,
1176 SQ. FT. PER FLOOR,
PERIMETER = 140′.

Figure 1-16 A simple rectangular house

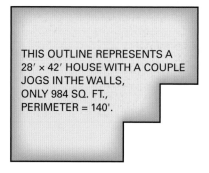

THIS OUTLINE REPRESENTS A
28′ × 42′ HOUSE WITH A COUPLE
JOGS IN THE WALLS,
ONLY 984 SQ. FT.,
PERIMETER = 140′.

Figure 1-17 A house with extra corners in the footprint has less square footage

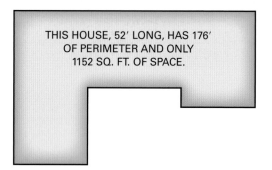

THIS HOUSE, 52′ LONG, HAS 176′
OF PERIMETER AND ONLY
1152 SQ. FT. OF SPACE.

Figure 1-18 This house has a much larger perimeter but less square footage than the house in Figure 1-16

Overall Dimensions

When considering material efficiency, keep the exterior dimensions in multiples of two. Sheet goods used on houses will easily fit into this pattern without generating unnecessary waste. For example, a house 32′ × 24′ will not cost significantly more than a house with dimensions of 31′ × 23′ (which is 55 sq. ft. smaller). This is partially due to the waste generated by the odd foot of material being discarded. Framing lumber is generally bought in 2′ increments, while sheet goods are most often 4′ × 8′.

Note: Some sheet goods are available in lengths up to 16′ and widths up to 5′ but are not as readily available. Engineered lumber, often used for joists, rafters, headers, etc., is commonly available in 1′ increments, or can be specially ordered to an exact length.

FOUNDATIONS

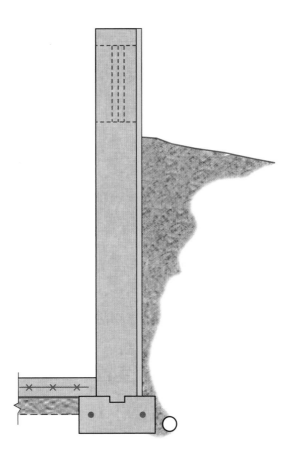

FOUNDATIONS

This chapter is intended as a brief overview of some common foundation types. It will aid in understanding the whole house package. This is *not* a construction guide for building footings and foundations.

Chapter Two Glossary

See Figure 2-1 for reference.

Footing—Supports the foundation walls. Generally at least twice as wide as the foundation wall, it provides a wide base to distribute the weight of the structure. It is most often located beneath the frost line.

Foundation—Most often a masonry support wall built on top of the footing. Generally, most of the foundation wall is located below ground level.

Frost line—The maximum depth beneath the ground surface that freezes during the cold months. This depth varies geographically (see Figure 2-6).

Pythagorean Theorem—A mathematical formula used to calculate the length of the hypotenuse of a right triangle. Carpenters use it in various ways, including keeping walls square to each other (see Figure 2-14).

Reinforcement bar—Abbreviated "rebar", or "bar." Lengths of steel bars that are placed at specific intervals, vertically and horizontally, inside of masonry walls/footings. Also can be used to connect footings to the foundation wall.

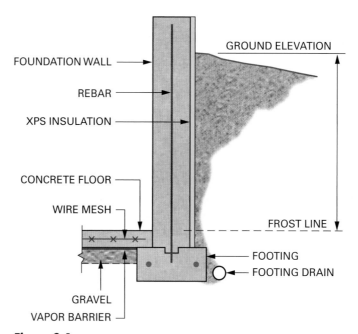

Figure 2-1 Foundation component terminology

FOOTING BASICS

Sizing Footings

Footing sizes are based upon building size/weight and soil type. There are some basic rules of thumb, but they should never be substituted for an engineered design.

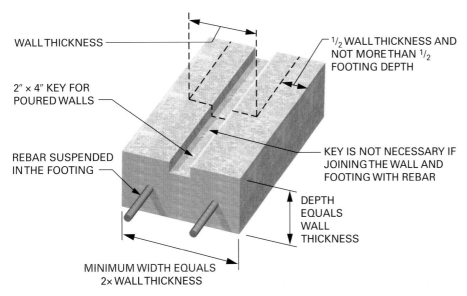

Figure 2-2 Typical footing for residential construction

Reinforcement Bar

Abbreviated "rebar," it is placed at specific intervals, vertically and horizontally, and helps to strengthen the footing and walls. It is also often used to connect the wall to the footing. Use of rebar in masonry walls and footings is generally a building code requirement. Rebar is generally sold with a number indicating its thickness. Each number refers to the number of eighths of an inch of thickness. For example, #5 rebar is 5/8″ thick, and #4 rebar is 1/2″ thick.

FOUNDATION TYPES

Slab on Grade

This option is popular in southern climates, where frost heaving issues do not exist, see Figure 2-3. It is possible to build slab-on-grade foundations in the North; however, there are some precautions that must be followed, such as preparing the bed with stone and properly insulating the slab and the area surrounding the slab.

Monolithic Slab

This has a thicker perimeter edge that also acts as a footing. The Monolithic slab is generally poured in combination with the floor. This can save time. See Figure 2-4.

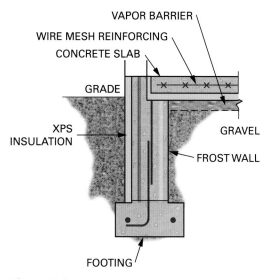

Figure 2-3 Slab on grade

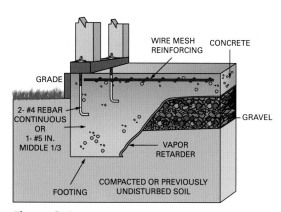

Figure 2-4 Monolithic slab

> *In northern climates, the foundation walls of the slab are typically insulated with Extruded Polystyrene Insulation. Building codes dictate the placement and amount. Some of these diagrams do not show the insulation.*

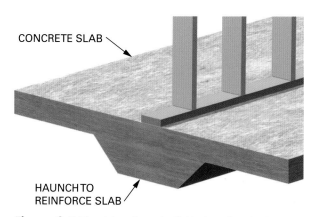

Figure 2-5 The slab is "haunched" (thickened) under load-bearing walls

> *Frost walls are less expensive than full foundations; however, there is no usable living space beneath the first floor.*

Frost Wall

A frost wall is a short wall that is constructed on top of the footing. The bottom of the footing must extend below the area that freezes in the winter, thus eliminating heaving (lifting) caused by the ground freezing. Figure 2-6 represents a map of the United States illustrating the areas of frost depth penetration.

Some foundations are built so that the interior of the frost wall is back-filled level, and an insulated slab-on-grade floor is poured (Figure 2-3). Others will have the top of the frost wall built 2′ or so above the surrounding grade. This allows the crawl space beneath the structure to be vented and accessed. Access to a crawl space is necessary to facilitate house maintenance. See Figure 2-7.

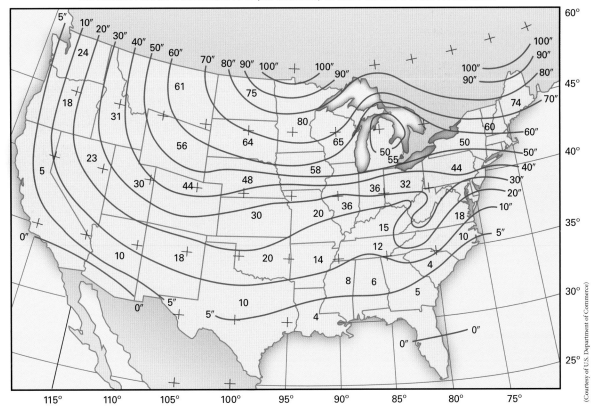

EXTREME FROST PENETRATION (IN INCHES) BASED UPON STATE AVERAGES

Figure 2-6 Frost line penetration in the United States

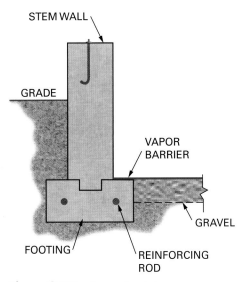

Figure 2-7 Crawl space foundation

Keep in mind that with any type of a slab, mechanicals such as plumbing supply and waste lines, in-floor heating, electrical conduit, and ductwork will all have to be accurately placed before the slab has been poured.

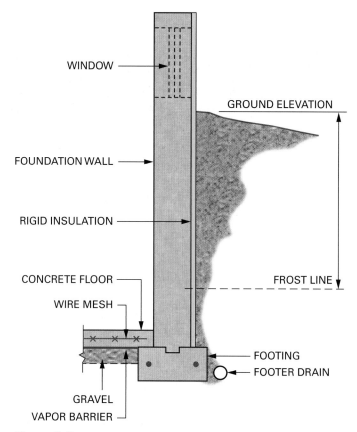

Figure 2-8 Full basement foundation

As with any foundation, drainage is important, but with a full-height basement (potential living space), attention to waterproofing and drainage are especially important.

Full Foundation

These foundations historically were very popular in northern climates, the reason being, in the North, a large amount of excavation must be completed in order to construct a footing beneath the frost line; and by digging a little deeper (and removing the dirt from the basement area), an extra floor level, the basement, can be easily constructed. See Figure 2-8. Conversely, in southern climates, where a monolithic slab on grade is often adequate (due to no risk of frost), a full foundation would add considerably to the cost of a house.

The first floor of a house with a full foundation is generally raised high enough above the grade that small windows can be placed in the foundation wall; this allows natural light into the basement area. Because of this, the floor of a basement with an 8′ ceiling may be only 6′ (or less) below the grade. See Figure 2-8.

Frost-Protected Shallow Foundation

This type of a foundation is sometimes known as an Alaskan Slab.

In recent years, there is more emphasis on reducing disturbance to the environment and on value engineering. For these reasons, and some code changes, the number of frost protected shallow foundations built is increasing.

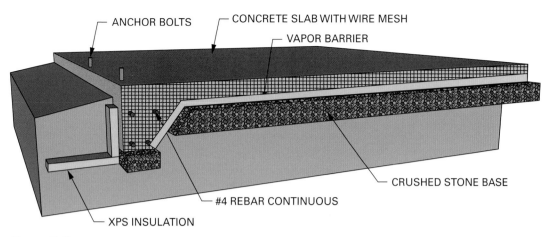

Figure 2-9 Frost-protected shallow foundation

Varying geographic regions have specifications for insulation thickness, distance below the surface for the insulation to be placed, and how far the insulation must extend outward from the foundation.

Even though the footing (commonly a monolithic slab) is not below frost level, special insulation details allow heat to be trapped from the earth to keep the foundation from freezing.

Notice in Figure 2-9 the XPS insulation extending outward from the slab several feet from the foundation.

Pressure-Treated Posts (Pole Barn)

While this may not qualify as a traditional foundation, this is a popular and inexpensive alternative to a masonry foundation. Many outbuildings and single-story additions, porches, and decks are built on pressure-treated-posts. The posts are set on footings that are located beneath the frost line. The size and spacing of the posts depend on the size of the structure being built. The floor system can either be suspended on the posts above the ground level, or an independent slab on grade can be poured, but this type of slab does not carry the weight of the structure and may not need to be thickened at the edge.

Disadvantages of building on posts:

- Lower resale value of the home/structure.
- Possible longevity/durability issues.
- Susceptibility to insects.

Advantages of building on posts:

- Often these structures are built without the use of expensive excavation equipment.

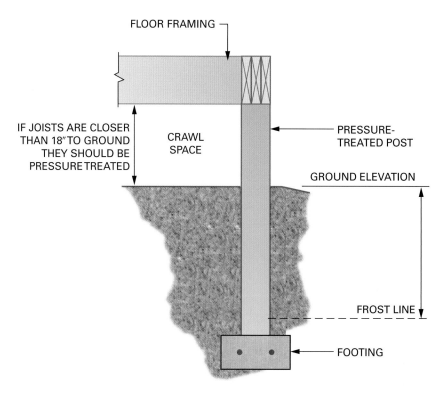

Figure 2-10 Post or pole barn construction

- Post holes can be dug by hand in the case of equipment inaccessibility to the job site.
- Minimal disturbance to the environment.

Other

Pressure-Treated Wooden Foundations These are constructed similarly to the walls of a house. Special details must be taken into consideration to prevent inward bowing, moisture problems, and insect issues (see Figure 2-11). These types of foundations generally cause the structure to have a lower resale value.

Insulated Concrete Forms (ICFs) (Not shown) These are becoming more and more popular. Special insulated block forms are stacked and braced, and concrete is then pumped inside them. The insulated blocks become a permanent part of the foundation.

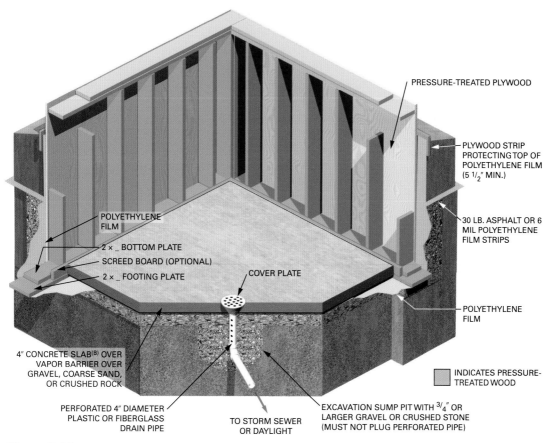

PRESSURE-TREATED PLYWOOD

PLYWOOD STRIP
PROTECTING TOP OF
POLYETHYLENE FILM
(5 $\frac{1}{2}$" MIN.)

30 LB. ASPHALT OR 6
MIL POLYETHYLENE
FILM STRIPS

POLYETHYLENE
FILM

POLYETHYLENE
FILM

2 × _ BOTTOM PLATE

SCREED BOARD (OPTIONAL)

2 × _ FOOTING PLATE

COVER PLATE

INDICATES PRESSURE-
TREATED WOOD

4" CONCRETE SLAB(B) OVER
VAPOR BARRIER OVER
GRAVEL, COARSE SAND,
OR CRUSHED ROCK

PERFORATED 4" DIAMETER
PLASTIC OR FIBERGLASS
DRAIN PIPE

TO STORM SEWER
OR DAYLIGHT

EXCAVATION SUMP PIT WITH $\frac{3}{4}$" OR
LARGER GRAVEL OR CRUSHED STONE
(MUST NOT PLUG PERFORATED PIPE)

Figure 2-11 Wood foundations are made with pressure-treated lumber

Figure 2-12 Pre-cast concrete foundations are delivered on
trucks and craned into place

Modular Foundations These are brought directly from the manufacturer to the job site and are set on a prepared bed of gravel. An advantage is that an entire Pre-cast foundation can be completed in a day (not including the excavation and site preparation). See Figure 2-12.

SQUARING AND CHECKING FOR SQUARE

Before starting construction it is good practice to check that the foundation has the proper dimensions and that it is square. Square refers to the corners being at a 90° angle to each other. The following processes will work for any square or rectangular structure.

Corner to Corner

The easiest way to confirm an area is square is to first measure the outside dimensions, both length and width; if dimensions are correct, then measure the diagonals of the corners. If the diagonal measurements are identical, then the corners are square; if not, the foundation is out of square. A minor adjustment may be possible while installing the sill plates (see Chapter 3, Figure 3-6).

This method of checking for square will not work if the opposite sides are different lengths, such as with a trapezoid shape.

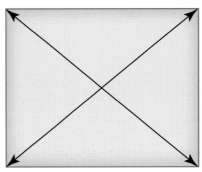

Figure 2-13 Measure from corner to corner to check for square

Pythagorean Theorem

Carpenters often call this the "3, 4, 5 rule" without knowing that it refers to the Pythagorean Theorem. Somewhere along the way, many carpenters learn that when one leg of a triangle measures 3′, the other leg 4′, and the hypotenuse 5′, then the two legs are square with each other (see Figure 2-14). This works with *any* ratio of 3, 4, and 5, such as 30′, 40′, and 50′ (multiplying each number by ten), or 6′, 8′, and 10′ (doubling each of the numbers).

The formula associated with the Pythagorean Theorem states that $\mathbf{a^2 + b^2 = c^2}$.

- "a" represents one leg of the right triangle.
- "b" represents the other leg of the triangle.
- "c" represents the hypotenuse (side opposite the right angle).

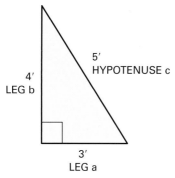

Figure 2-14 Using the
Pythagorean Theorem

For example, assume a = 3, b = 4, and we are be trying to find c. By plugging in the numbers,

$$3^2 + 4^2 = c^2$$

Therefore,

$$9 + 16 = c^2$$

Therefore,

$$25 = c^2$$

By taking the square root of 25, we find that c = 5. This process will work with any set of numbers. See another example in Chapter 2 Appendix.

Some houses have many angles and can be broken into smaller rectangles. Some diagonals will need to be calculated and measured in order to check for square. Some of the possible diagonals that can be checked on an odd-shaped building are shown in Figure 2-15.

The Pythagorean Theorem is a very powerful mathematical formula that can be used by carpenters many different ways. It is well worth the time to learn this concept, as it will increase the accuracy of a building project and save time.

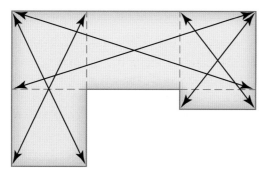

Figure 2-15 Some of the possible diagonal measurements that can be checked

If building with odd (non-square) angles, sometimes it is easier to square a section of the area and then measure from the square points, such as point E in Figure 2-16.

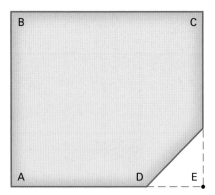

Figure 2-16 Squaring an odd shape

SILLS, GIRDERS, AND POSTS

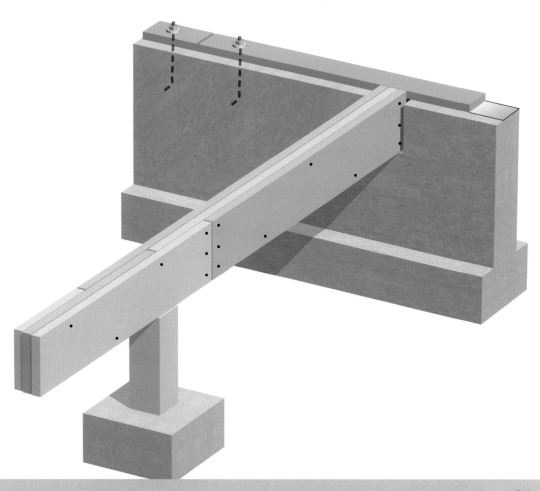

SILLS, GIRDERS, AND POSTS

This is most commonly the starting place for a carpenter.

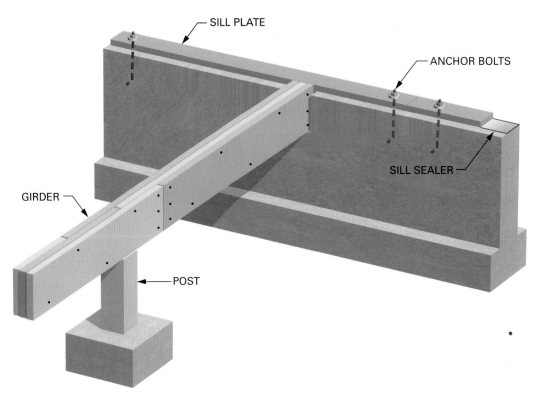

Figure 3-1 The starting components for a floor system

Chapter Three Glossary

Refer to Figure 3-1 for locations of components described below.

Anchor bolt—A special fastener connecting the sill plate to the foundation wall.

Girder—A load-bearing horizontal framing member (beam). In this case it supports the load from the floor above it.

Post—A vertical support member that often helps to carry the girder.

Sill plate—Fastened to the foundation by anchor bolts, sills provide a wood surface to nail to. Generally a pressure-treated 2 × 6.

Sill sealer—A thin layer of insulation fitting between the sill plate and the top of the foundation wall. As building weight is added, sill sealer compresses and seals out drafts.

SILL PLATES

Installation of the sill plates is often the starting point for a carpenter. Typically a layer of sill sealer is first placed on top of the foundation wall, and then the sill plates are placed according to the procedure in Figure 3-2.

STEP **1** SNAP CHALK LINE ON FOUNDATION WALL.

STEP **2** ALIGN SILL PLATE AGAINST ANCHOR BOLTS AND PARALLEL TO CHALK LINE.

STEP **3** SQUARE LINES ON SILL FROM BOTH SIDES OF EACH ANCHOR BOLT.

STEP **4** MEASURE EACH BOLT DISTANCE FROM CHALK LINE AND TRANSFER TO SILL.

STEP **5** DRILL HOLES APPROXIMATELY 1/8" LARGER THAN BOLT DIAMETER.

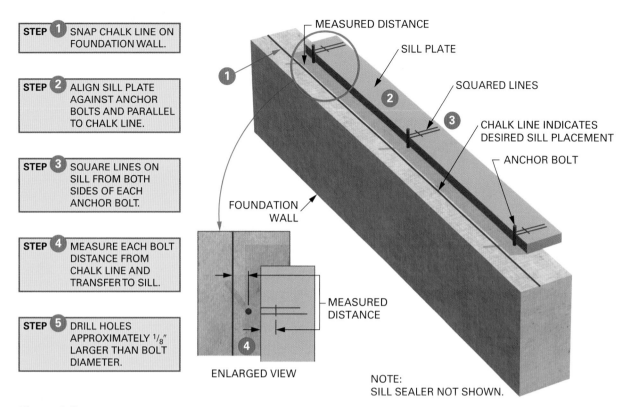

Figure 3-2 Installing a sill plate on a foundation using anchor bolts

Sill Anchors

There are various types of anchors available to fasten sill plates to the foundation wall; see Figure 3-3 for some common variations.

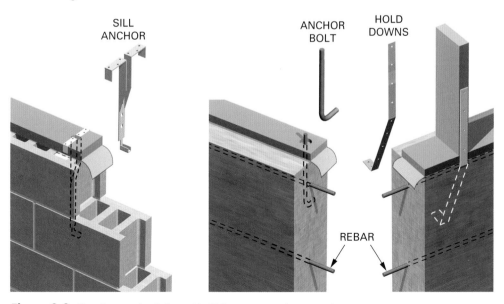

Figure 3-3 Sill anchors, anchor bolts, and hold downs connect frame members to concrete

Variations on Sill Placement

Sills can be placed flush with the foundation wall, set back slightly, or even overhang slightly depending on the exterior surface material (check local codes for the amount of overhang allowed). See Figure 3-4.

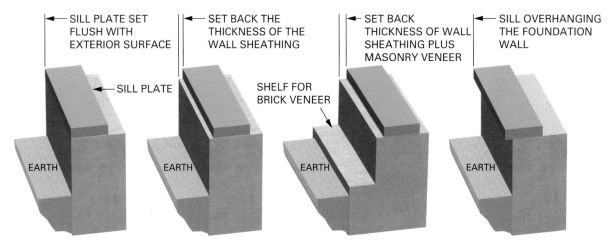

Figure 3-4 A sill plate may be placed differently depending on the given situation

Generally a single sill plate is adequate; however, adding a second sill plate is an easy way to raise the joists another 1½″. This may be necessary to increase the headroom in a basement.

> *Anchors must be placed between 4″ and 12″ from the end of the sill plates and not farther than 6′ apart elsewhere.*

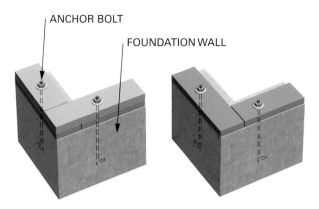

Figure 3-5 Double sill (left) and single sill (right)

Adding Anchors Where None Are Present

Occasionally, due to an oversight or when remodeling, an anchor may need to be added. Various anchoring hardware is available to make this process possible. One such product is an epoxy system, where a hole is drilled for the anchor and a special epoxy adhesive is injected into the hole, hardening around the anchor bolt. Other products include different types of expandable anchors, where an anchor made of several parts literally expands as it is tightened.

Adjusting for an Out-of-Square Foundation

Before starting to build, the carpenter should always check the foundation dimensions for accuracy and for squareness (see "Squaring and Checking for Square" in Chapter 2). In the event the foundation is slightly out of square, corrections should be made before proceeding further. Minor corrections are achieved by hanging the sill over the edges of the foundation at the opposite corners. This will aid in an attempt to create 90-degree angles. Obviously there are limitations to the amount the sill plate can overhang, check local codes.

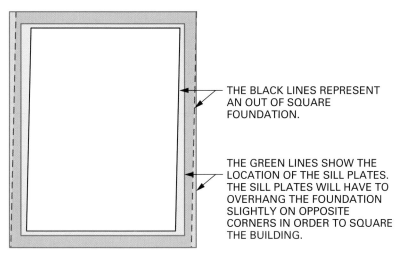

THE BLACK LINES REPRESENT AN OUT OF SQUARE FOUNDATION.

THE GREEN LINES SHOW THE LOCATION OF THE SILL PLATES. THE SILL PLATES WILL HAVE TO OVERHANG THE FOUNDATION SLIGHTLY ON OPPOSITE CORNERS IN ORDER TO SQUARE THE BUILDING.

Figure 3-6 Correcting an out-of-square foundation

GIRDERS

Girders act as intermediate supports for the joists. They may also carry weight from other sources above them. Girders can be framed flush with the joists (see Figure 3-7). They can be hung from the foundation wall with special hardware, or

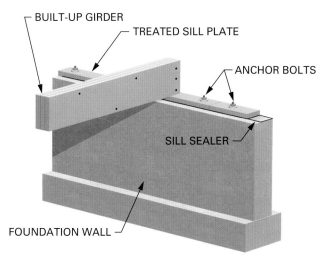

BUILT-UP GIRDER

TREATED SILL PLATE

ANCHOR BOLTS

SILL SEALER

FOUNDATION WALL

Figure 3-7 Flush girder

be placed on a support pilaster (Figure 3-8) integrated into the foundation wall (flush or dropped). Girders can also be "dropped" by placing them into a notch in the foundation wall called a beam pocket (Figures 3-9 and 3-10). When girders are dropped, the joists typically rest directly on top of them.

Girder Hardware

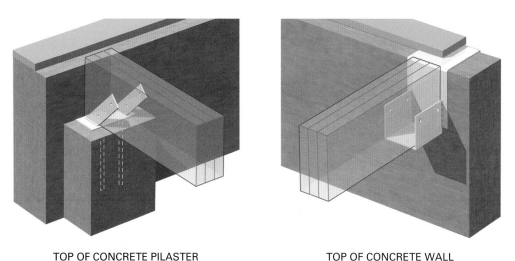

TOP OF CONCRETE PILASTER   TOP OF CONCRETE WALL

Figure 3-8 Girder and beam seats provide support from concrete walls

Girder Pocket

During construction of the foundation, a pocket is formed within the wall, eliminating the need for special hangers. Note the minimum clearance/size requirements shown in Figures 3-9 and 3-10.

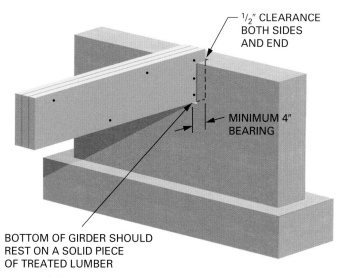

$^1/_2$" CLEARANCE
BOTH SIDES
AND END

MINIMUM 4"
BEARING

BOTTOM OF GIRDER SHOULD
REST ON A SOLID PIECE
OF TREATED LUMBER

Figure 3-9 A girder pocket of a foundation wall should be large enough to provide air space around the end and sides of the girder

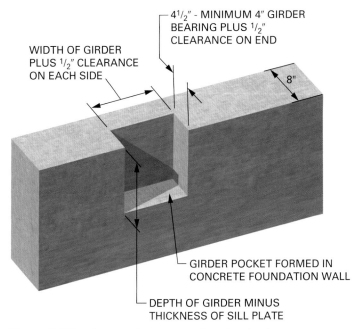

WIDTH OF GIRDER
PLUS $1/2$" CLEARANCE
ON EACH SIDE

$4^1/2$" - MINIMUM 4" GIRDER
BEARING PLUS $1/2$"
CLEARANCE ON END

8"

GIRDER POCKET FORMED IN
CONCRETE FOUNDATION WALL

DEPTH OF GIRDER MINUS
THICKNESS OF SILL PLATE

Figure 3-10 Girder pocket locations are indicated on foundation plans

Types of Girders

Figure 3-11 illustrates some girders that can be made of different types of materials, such as solid wood, built-up girders (from lumber such as 2 × 12's), steel, and engineered lumber (see Chapter 4, "Engineered Lumber Options").

Built-up Girder A built-up girder, shown in Figure 3-12, has the advantage of being able to be built in place one piece at a time. This may eliminate some of the manpower or equipment necessary to place a heavy girder.

There is a specific process to constructing a girder; this includes sizing of members, locating splices, and nail/bolt spacing. In Chapter 3 Appendix, there are tables and diagrams that help determine girder sizes appropriate to the load and span. Built-up girders should be designed by a qualified professional.

POSTS

Posts are generally needed to support girders at regular intervals.

Post Types and Spacing

Solid 6″ × 6″, pressure-treated wood posts are often used, and they stand on a special base connector that rests on top of the concrete floor or footing. 6 × 6's are commonly used to replace the temporary posts that are used during construction (see Figure 3-13).

Concrete-filled steel posts, when used, rest directly on the center of a footing, and the floor is placed around them. These posts have to be cut to an exact length before positioning (see Figures 3-14 and 3-15).

SOLID LUMBER JOISTS

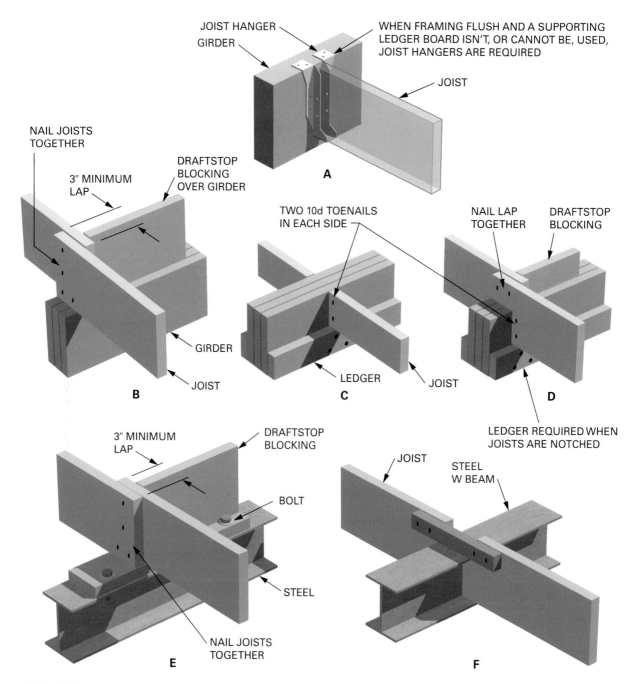

NOTE: *DUE TO ALL OF THE DIFFERENT TYPES OF GIRDERS AND THE IMPORTANCE OF THE STRUCTURAL ASPECTS, ALWAYS FOLLOW THE PLAN'S SPECIFICATIONS.
*SOME BUILDING CODES MAY REQUIRE THAT A BUILT-UP GIRDER BE BOLTED AND/OR GLUED TOGETHER.
*IN ORDER TO SAVE HEADROOM IN THE FLOOR BELOW, CONSIDER, IF POSSIBLE, USING A FLUSH GIRDER.

Figure 3-11 Various possible framing details at a girder

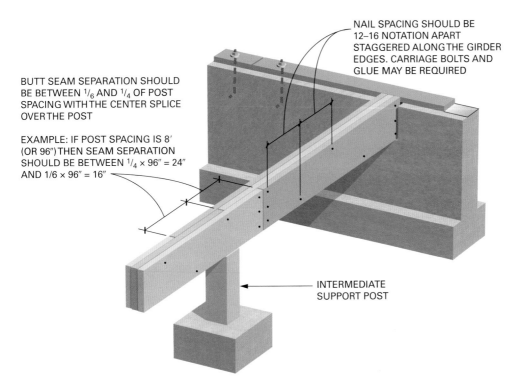

NAIL SPACING SHOULD BE 12–16 NOTATION APART STAGGERED ALONG THE GIRDER EDGES. CARRIAGE BOLTS AND GLUE MAY BE REQUIRED

BUTT SEAM SEPARATION SHOULD BE BETWEEN $1/6$ AND $1/4$ OF POST SPACING WITH THE CENTER SPLICE OVER THE POST

EXAMPLE: IF POST SPACING IS 8' (OR 96") THEN SEAM SEPARATION SHOULD BE BETWEEN $1/4 \times 96" = 24"$ AND $1/6 \times 96" = 16"$

INTERMEDIATE SUPPORT POST

NOTE: CARRIAGE BOLTS MAY BE REQUIRED WHEN CONSTRUCTING A BUILT-UP GIRDER.

Figure 3-12 Spacing of fasteners and seams of a built-up girder made with dimension lumber

There are many types of hardware available for joist-girder-post connections. Only a few are shown here.

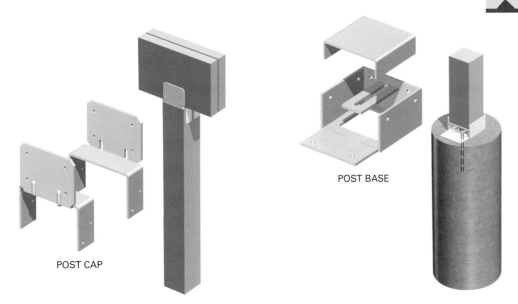

POST CAP

POST BASE

Figure 3-13 Special hardware helps to fasten tops and bottoms of posts and columns

Notice the footing the post rests on. A load bearing post should not be placed directly on a concrete floor unless there is a footing beneath to spread the load.

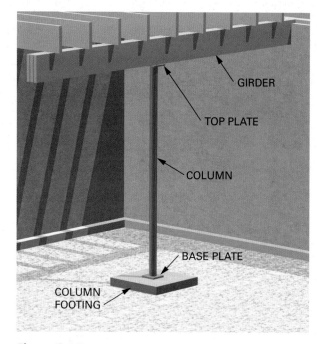

Figure 3-14 Typical column supporting a girder

STEP ①	WRAP WIDE SHEET OF PAPER AROUND COLUMN.
STEP ②	KEEP EDGES OF PAPER EVEN.
STEP ③	MARK AROUND COLUMN ALONG EDGE OF PAPER.
STEP ④	USING A SAW WITH AN APPROPRIATE BLADE, CUT THE POST.

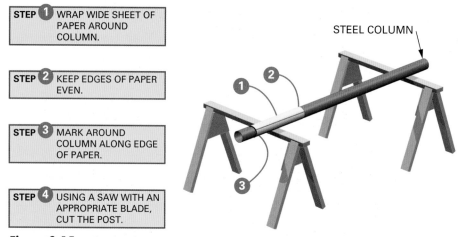

Figure 3-15 Marking a square line on a column

Telescoping posts (not shown) are most often for **temporary** use (during early construction) and are replaced later with something more substantial.

FLOOR FRAMING

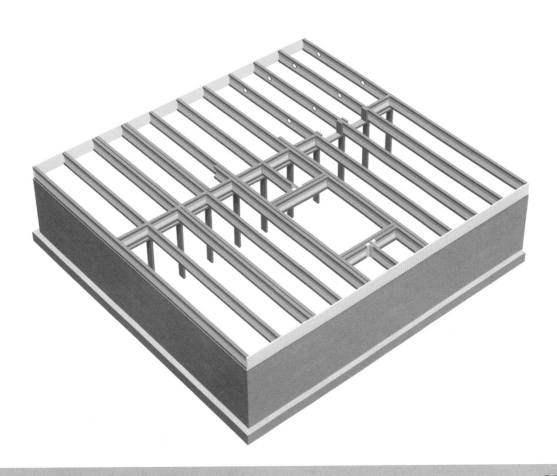

FLOOR FRAMING

Framing terminology and symbols may vary from region to region. In this book the terms shown in Figure 4-1 will be used.

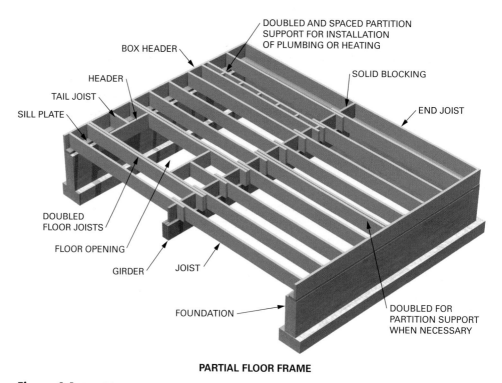

PARTIAL FLOOR FRAME

Figure 4-1 Typical framing components of a floor system

Chapter Four Glossary

Blocking—Short solid wood pieces placed between joists. Solid blocking can be used as bridging and can double as a draft stop over the top of a girder.

Box header (band joist, rim joist)—Framing member at the edge of the building that the other joists are nailed to.

Bridging—Placed between the joists from the top of one joist to the bottom of the next, two per joist cavity; the two pieces form an "X" shape and help with lateral stability. Bridging is not always necessary depending on joist size and span.

Centerline—Measurements on plans to the center of an opening or the center of a partition, called centerline measurements—not to be confused with *OC*.

Header—Framing member designed to carry a heavier load. Often made of two or more pieces fastened together.

Joist—Floor support members placed at regular intervals, such as 16″ on center (16 OC). These are designated with an "X" in layouts.

Layout—Locating the placement of framing members.

On center—Abbreviated "OC." Refers to the regular spacing of the framing members, measured from the center of one member to the center of the next—not to be confused with *centerline.*

Sub-floor—Most often constructed with sheet goods such as plywood or Oriented Strand Board (OSB). It is secured directly to the floor joists.

Tail joist—Also called a cripple joist, a short OC joist usually found butting up against a header opposite an opening. Designated with a "C" in layouts.

Trimmer—Full-length joist that is placed next to an opening such as a stairwell. The trimmer is placed next to a regular joist to help it carry the load. Some call the two joists "double trimmers." Designated with a "T" in layouts.

This chapter will focus most closely on framing with solid wood. However, along the way, there will be some comments on engineered lumber. Regardless, whether framing with engineered lumber or dimension lumber, concepts of layout and assembly are similar. Manufacturers of engineered lumber products will generally provide a pamphlet with their products to aid in properly cutting, boring, and fastening them.

Lumber today is sold mostly in nominal sizes, not actual sizes. In other words, a 2 × 4 does not measure 2″ by 4″; it actually measures 1½″ × 3½″. The length is always given in feet and is accurate. This table will help to understand the lumber sizes.

Actual Size	Nominal Size	Actual Size	Nominal Size
2 × 4	1½″ × 3½″	1 × 4	¾″ × 3½″
2 × 6	1½″ × 5½″	1 × 6	¾″ × 5½″
2 × 8	1½″ × 7¼″	1 × 8	¾″ × 7¼″
2 × 10	1½″ × 9¼″	1 × 10	¾″ × 9¼″
2 × 12	1½″ × 11¼″	1 × 12	¾″ × 11¼″

Lumber can be obtained in some places "rough cut." In this state, it will not have been planed smooth, a rough-cut 2 × 4 actually measures 2″ × 4″. Most carpenters today prefer to build with nominal-sized lumber.

For the purpose of this book, we will assume the use of nominal-sized lumber.

METHODS OF FLOOR ASSEMBLY

Shown below are two common methods, each has advantages and disadvantages. Either method is good and can be equally accurate. What is important is choosing a method that one is comfortable with.

Method 1

1. Strike a chalk line on the sill plates 1½" from the outside edge; if not using standard dimension lumber, this distance will be different. This will locate the end of the joists and the inside edge of the box header. Next place layout marks on the sill plates (covered later in this chapter).

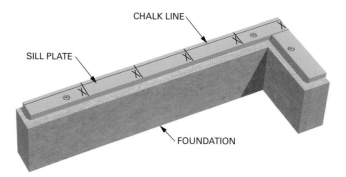

Figure 4-2 Place layout marks on the sill plate

2. Nail the joists to the sill at the chalk line and on the layout marks.

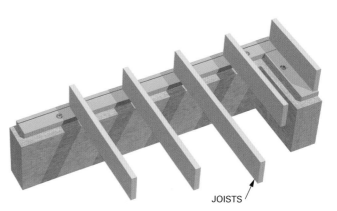

Figure 4-3 Set joists to the chalk line

3. Add the box header, squaring the joists to the box header while nailing the box header to the joists and to the sill.

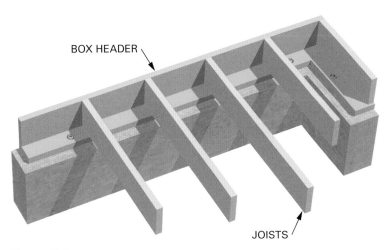

Figure 4-4 Nail box header square to joists

Method 2

1. Place layout marks on the *box header.*

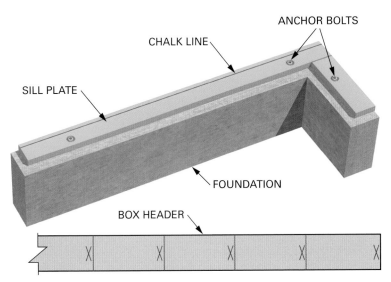

Figure 4-5 Place layout lines on the box header

2. Strike a chalk line on the sill plates to locate the inner edge of the box header (generally 1½″ in from sill edge).

3. Nail the box header to the sill at the chalk line.

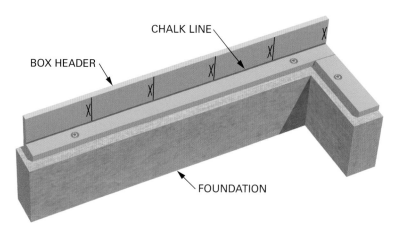

Figure 4-6 Nail box header to sill at the chalk line

4. Place the joists to the layout lines on the box header, nailing the box header to the joists and the joists to the sill.

NOTE

There are variations of each of these methods that carpenters may choose to use.

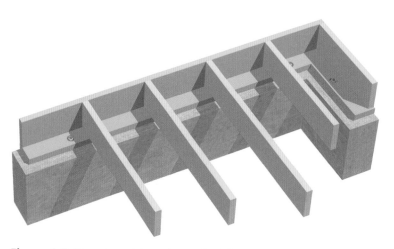

Figure 4-7 Align joists with layout lines and nail through box header into joists

BASIC DESCRIPTION OF LAYOUT

Below is a quick overview of basic layout; a more detailed description will follow. The floor plan depicted in Figure 4-8 coincides with the framed floor in Figure 4-12.

> *Notice how the locations of the partitions influence the joist placement below. Walls/partitions must be taken into consideration when framing the floor.*

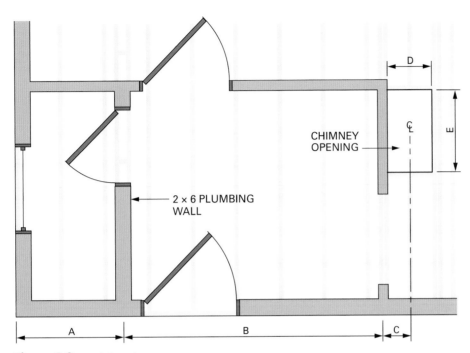

Figure 4-8 Partial floor plan

Figure 4-9 Step 1: Measure and mark chimney opening centerline, then measure and mark double joists on each side of the opening

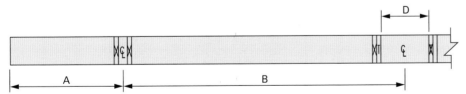

Figure 4-10 Step 2: Measure and mark partition centerline, then measure width of partition and mark a joist on both sides of the partition. The partition on the left is a plumbing wall and a space between the joists is necessary to make room for piping

Figure 4-11 Step 3: Mark the first OC joist by deducting one-half the joist thickness (generally 3/4″) from the desired OC spacing, then mark remaining joists according to the OC spacing. Any joists that fall between Trimmers ("T") are marked with a "C"

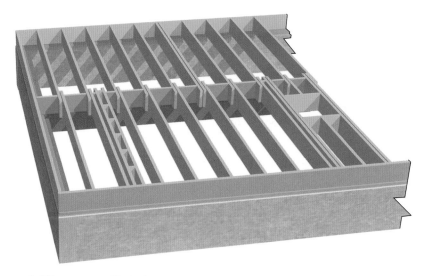

Figure 4-12 Step 4: Assemble the framing. Shown above are typical components of a framed floor system

DETAILED DESCRIPTION OF LAYOUT

Due to the extreme importance of an accurate layout, a detailed explanation follows on the finer points and reasoning behind layout procedures.

Crowning

Select the straightest lumber for the box header. Crowning is accomplished by sighting down the length of a board and placing an indicator mark ∧ pointing to the edge of the board that crowns upward (see Figure 4-13). This process is not necessary when using engineered lumber.

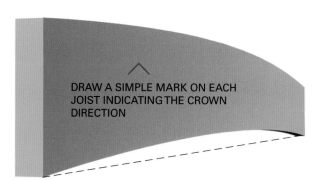

Figure 4-13 Crowning a board

Horizontal framing members, such as joists, are placed with the crowns up. The straightest lumber becomes the box header and will be placed directly on the sill plate; it should be in full contact with the sill. This way, as building weight is added, it will *not* settle.

> **TIP**
>
> *While selecting box header material, crown all other joist lumber. This will save time during assembly of the floor system.*

Locations of Centerlines

Centerline measurements are most often dimensioned from the building line (from the framing at the outer edge of the building, not including the wall sheathing). When framing a floor, *centerline* dimensions may be shown either to the center of a partition or to the center of an opening. These centerlines are indicated by a "C" and "L" superimposed together (℄) and are the **first marks** to be placed on the box header (see Figure 4-16). The long part of the "L" should be the line marking the distance measured.

TIP

Remember, dimensions on plans do not include the sheathing, only the framing.

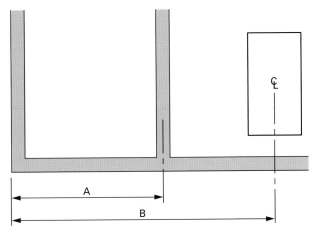

Figure 4-14 Centerline dimensions can be shown to the center of a partition (A) or to the center of an opening (B)

Locations of Openings

Openings in floors are framed for stairwells, chimneys, and laundry chutes. Occasionally, it may be necessary to frame a chase for mechanicals, such as large ductwork or plumbing that will not fit into a partition cavity. Openings can be dimensioned in three different ways on plans (see Figure 4-15).

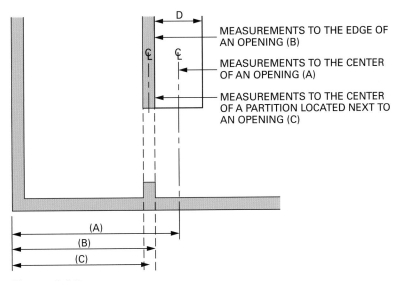

MEASUREMENTS TO THE EDGE OF AN OPENING (B)

MEASUREMENTS TO THE CENTER OF AN OPENING (A)

MEASUREMENTS TO THE CENTER OF A PARTITION LOCATED NEXT TO AN OPENING (C)

Figure 4-15 Three ways to dimension openings

Dimension "D" (Figure 4-15) represents the rough opening width.

Methods of Marking Openings on the Box Header

Method A. Find the measurement on the plans, mark the centerline, then measure ½ of the rough opening (RO) dimension in either direction from the centerline. Mark two joists on each side of the opening.

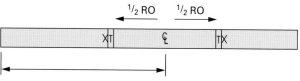

Figure 4-16 Method A. Measurement to the center of an opening

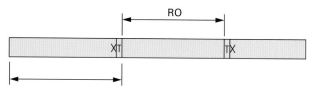

Figure 4-17 Method B. Measurement to the edge of an opening

Method B. Using dimensions found on the plans, mark the distance to the edge of the opening. Next, mark the rough opening width found on the plans. As given above, mark two joists on either side of the opening.

Openings often require two or more joists on each side; the extra joists help to carry the load from the header located at either end of the opening. Even though doubled joists are identical, this book will refer to the inner joist as a trimmer and designate it with a "T" symbol, and the outer joist right next to the "T" will be designated "X." Using the "T" in combination with "X" alerts the person who is assembling the floor that there is an opening between the two joists marked "T."

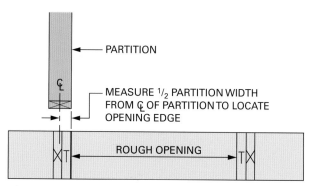

Figure 4-18 Method C. Measurement to the center of a partition next to an opening

Method C. Occasionally, openings are determined by the locations of partitions. Locate the partition centerline dimension from the plans and mark the box header accordingly. Find the edge of the partition by measuring over ½ the partition width from the centerline. This measurement will depend on the partition stud size, generally a 2 × 4. From this line, measure the rough opening width to locate the other side of the opening. As above, mark the double joists on each side of the opening.

Locating and Reinforcing Parallel Partitions

Parallel partitions are partitions that run parallel to the floor joists and, even though they are typically non–load bearing, they need extra support. Occasionally it is adequate to simply double a joist directly beneath the partition (Figure 4-19A). Other times it is necessary to leave the space directly below the partition open so mechanicals (plumbing, electrical, etc.) can pass up through the floor and into the partition's cavity (Figure 4-19B). In remodeling situations, or if there are last-minute changes, it may be easiest to simply add some horizontal blocking between adjacent joists to stiffen up the floor (Figure 4-19C). The blocking is placed at regular intervals, such as 16 OC.

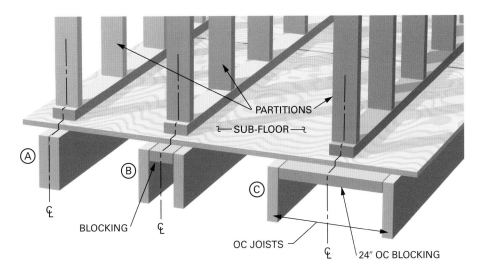

Figure 4-19 Methods of supporting parallel partitions

Most load-bearing partitions are located over the tops of girders and run perpendicular to the joists. However, if the partitions shown in Figure 4-19 are load bearing, additional support will be required, such as support posts beneath the joists.

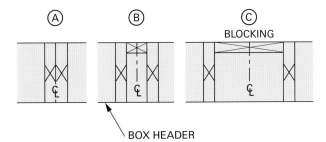

Figure 4-20 Box header depicting joist layout marks supporting parallel partitions (refer to Figure 4-19)

Location of the On-Center Joists

On-center joists are laid out from the building line. Commonly used layout spacing is 16″ on center (16 OC). This means that the distance from the center of one joist to the center of the next is 16″. There is one exception. **The first joist is offset ¾″**, half the thickness of a 1½″ (2×) joist; see Figure 4-21. This ensures the *center* of subsequent joists will be located at multiples of 16″. In Figure 4-21 the edge of the first joist is located 15¼″ from the building line, and the second joist is located 16″ from the first, or 31¼″ from the building line. Following that the edge of each joist is located at 47¼″, 63¼″, 79¼″, 95¼″, 111¼″, and so forth… notice the pattern. This allows one end of an 8′ sheet of sub-floor material to be placed flush with the building line, and the other end will land on the *center* of the joist 8 feet away. By landing on the center, this leaves room for the next sheet to be fastened to the same joist.

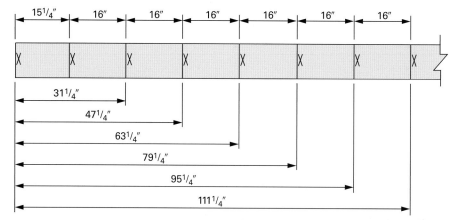

Figure 4-21 16″ On-center layout

Variations of On-Center Framing

Many floor systems are built with 24 OC spacing. If using 2× material (1½″ thick), the first joist is located 23¼″ from the building line and the next is 24″ from the first, which is 47¼″ from the building line, see Figure 4-23. The same concepts can be utilized to frame 12 OC, 19.2 OC, 32 OC, and other OC spacings.

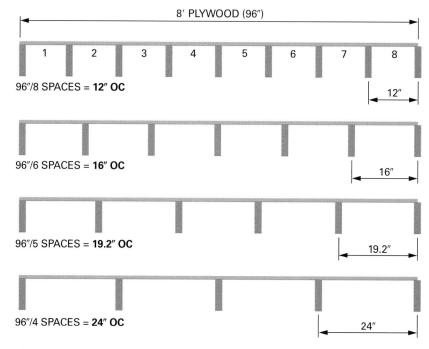

Figure 4-22 Some of the possible on-center spacing intervals

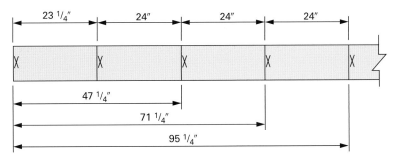

Figure 4-23 24 OC layout

Order of Assembly

Framing openings such as chimneys and stairwells in a specific order can save time. Follow the procedure in Figure 4-24 and consider the assembly order. The suggested order ensures *end nailing* is possible instead of toe nailing.

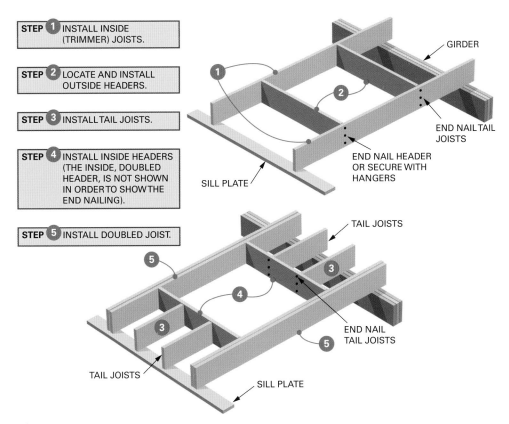

STEP 1 INSTALL INSIDE (TRIMMER) JOISTS.

STEP 2 LOCATE AND INSTALL OUTSIDE HEADERS.

STEP 3 INSTALL TAIL JOISTS.

STEP 4 INSTALL INSIDE HEADERS (THE INSIDE, DOUBLED HEADER, IS NOT SHOWN IN ORDER TO SHOW THE END NAILING).

STEP 5 INSTALL DOUBLED JOIST.

Figure 4-24 Installing framing members around a floor opening

FRAMING PROCEDURES AT THE SILL AND OPTIONS AT THE GIRDER

Flush at Girder with Joist Hangers

There may be headroom issues that will not permit the joists to sit on top of the girder. If joists are framed flush with the girder, the girder will be higher from the floor than if built directly under the joists. This method takes more time, accurate cuts are needed, and approved hangers must be used at the girder (see Figure 4-25).

When header joist span is over 4', the trimmer and header joists must be doubled. When header joists are 6' or longer, secure with joist hangers. When tail joists are longer than 12', connect with joist hangers (Figure 4-25).

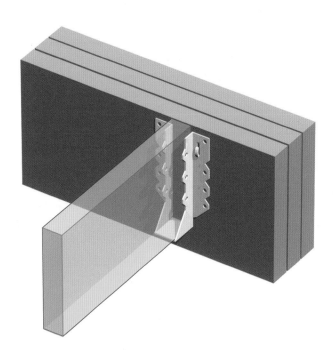

NOTE: WHEN SECURING JOIST HANGERS, USE ALL
NAIL HOLES PROVIDED AND USE
THE APPROPRIATE NAILS

Figure 4-25 Hangers are used to support joists and beams

Flush at Girder with Ledger Support

Codes generally require a minimum of a 2" × 2" ledger support, whenever a ledger support is used.

A ledger can save headroom and eliminate the need for joist hangers in most cases. This method is only possible if the girder is larger than the joists, or if joists are raised and notched.

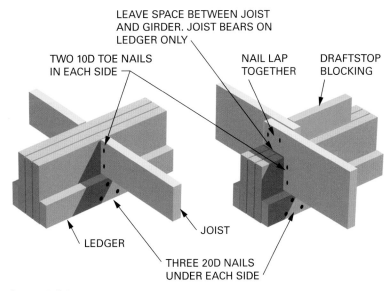

LEAVE SPACE BETWEEN JOIST
AND GIRDER. JOIST BEARS ON
LEDGER ONLY

TWO 10D TOE NAILS
IN EACH SIDE

NAIL LAP
TOGETHER

DRAFTSTOP
BLOCKING

JOIST

LEDGER

THREE 20D NAILS
UNDER EACH SIDE

Figure 4-26 Ledger supports can eliminate hangers

Framing Straight through (over) the Girder

This method is often used when framing with engineered lumber. Engineered lumber can be manufactured to long lengths and it is possible to span large openings. However, these joists may still need additional support at mid-span (see Figure 4-27).

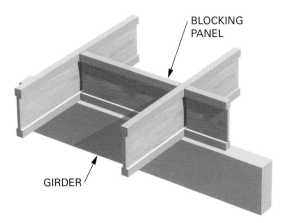

BLOCKING
PANEL

GIRDER

Figure 4-27 Joists framed straight through at the girder

Offset at the Girder

When framing with dimension lumber, it is unusual to be able to span an entire house foundation. Instead of butting the pieces at the center of the span, it is quickest to *offset* the joists on one side of the building by 1½″ so that joists sit side by side and overlap at the girder. This method saves time by not having to cut the joists (see Figure 4-28).

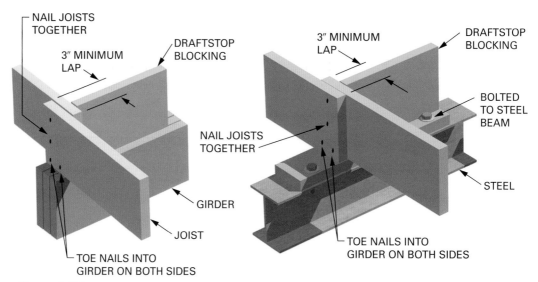

Figure 4-28 Joists offset at the girder

Offset Layout

The layout process on each side of the building is the same with the exception that on one side (the left in Figure 4-29), the layout marks "X" are placed on the opposite side of the layout line. This ensures a 1½" offset. Layout marks must also be placed on the girder; however, there the marks will be placed on both sides of the layout line and offset toward each corresponding side.

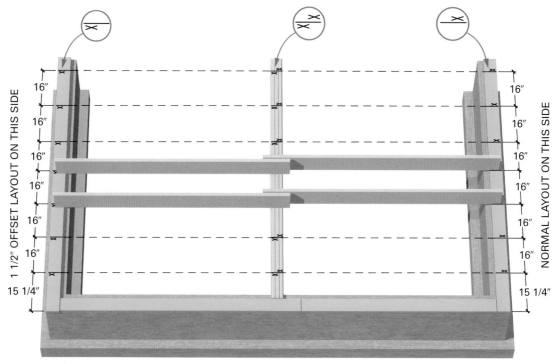

Figure 4-29 Layout marks are on opposite sides of the layout lines on each side of the building

BRIDGING/BLOCKING/STRAPPING

Solid Blocking at the Girder

Whether joists are offset at the girder (see Figures 4-28 and 4-29) or are framed straight through (Figure 4-27), solid blocking at the girder is necessary. The blocking serves two functions. It stabilizes the joists laterally, and in the event of a fire, it serves as a draftstop.

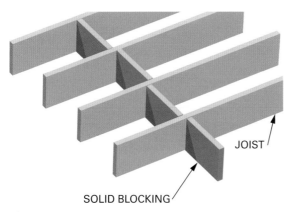

Figure 4-30 Solid blocking

Metal or Wood Bridging

Bridging is often required as part of the floor system (check local code); the bridging helps to stabilize the joists by connecting them together.

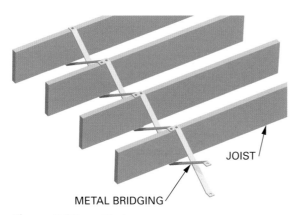

Figure 4-31 Metal bridging

> IRC code requires bridging when solid wood joists are larger than 2 × 12 and the span is greater than 8′. Follow manufacturer recommendations for engineered lumber such as I-joists.

Today, metal bridging has all but replaced the former wood bridging; however, there are some who still use wood bridging.

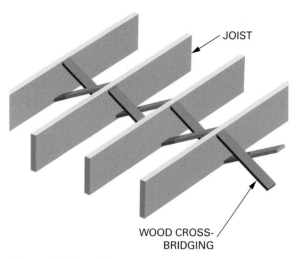

Figure 4-32 Wood bridging

Strapping

Strapping the bottoms of joists with either metal straps or a continuous line of wood 1×3's is another alternative. However, continuous strapping may get in the way if the ceiling is to be finished.

Making Wood Bridging

Bridging can be spaced a maximum of 8' apart per continuous row and is applied before the sub-floor is installed. If the joist span is less than 16', find the center and a snap a chalk line. Next, nail only the top of the bridging before applying the sub-floor. Later, after the rest of the house is framed, and before the bottoms of the joists are covered, attach the bottom of the bridging. Following is an easy method of layout and cutting of wood bridging.

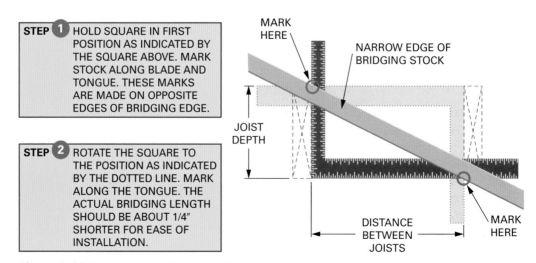

STEP 1 HOLD SQUARE IN FIRST POSITION AS INDICATED BY THE SQUARE ABOVE. MARK STOCK ALONG BLADE AND TONGUE. THESE MARKS ARE MADE ON OPPOSITE EDGES OF BRIDGING EDGE.

STEP 2 ROTATE THE SQUARE TO THE POSITION AS INDICATED BY THE DOTTED LINE. MARK ALONG THE TONGUE. THE ACTUAL BRIDGING LENGTH SHOULD BE ABOUT 1/4" SHORTER FOR EASE OF INSTALLATION.

Figure 4-33 Laying out cross-bridging using a framing square

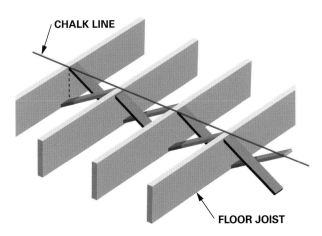

Figure 4-34 Aligning bridging

NOTCHING AND BORING HOLES IN JOISTS

There are specific locations and sizes of notches and holes that can be bored in the joists. Drilling or notching at the wrong location or notching too deep can cause joists to fail. Specifications may vary with different types of engineered lumber. The specifications below pertain to nominal 2 × lumber (1½″ thick).

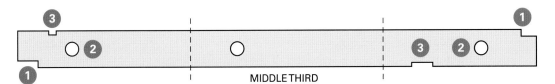

Figure 4-35 Notching and boring limits

1. Notches on ends of joists cannot exceed ¼ of the joist depth.
2. Holes in joists cannot be within 2″ of the top or bottom edge and the diameter cannot be more than ⅓ the depth of the joist.
3. Notches in the top or bottom edges cannot exceed ⅙ the joist depth and cannot be in the middle third of the joist.

Dead vs. Live Loads for Floor Systems

When referring to floor weight, the terms "dead load" and "live load" are used. Dead load simply means the weight of the floor material and other components, such as walls, that contribute weight to the floor. Live load refers to weight that can be moved or changed from time to time, such as people and furniture, or, when referring to roof loads, snow. There are specific building codes that are followed when constructing houses; these codes incorporate a margin of safety. Span tables that rate different species of wood, length of span, on-center spacing, and grade of wood can be found in the Chapter 4 Appendix. Use of these tables may vary from region to region, so always consult with a qualified professional when designing for a specific load. Much more extensive tables can be found in the IRC manual.

Hardware

Different types of hardware and braces can help during construction. Some types make a carpenter's job easier. Other times, building codes require specific hardware. An example of this is the special hardware that connects foundation walls to the sill plates and upward to the framed walls. Such hardware is required to combat forces that may occur in earthquake-prone areas or in high-wind areas. Be aware that hardware requirements vary regionally.

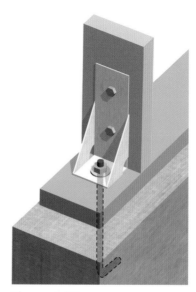

Figure 4-36 Example of specialized hardware

TIP

The second row can be started with either a half-sheet or the cut-off sheet from the end of the first row. No two consecutive rows should break on the same joist. This overlapping gives strength to the sub-floor. Always follow manufacturer recommendations regarding OC spacing between sheets.

SUB-FLOOR APPLICATION

See Figure 4-37 for the sub-floor application steps.

- **Step 1**—Strike a chalk line the length of the building on the tops of the joists, 48″ from the building line (outer edge). Apply construction adhesive to the top of the joists (up to the chalk line).
- **Step 2**—Start the first row with a full sheet, aligning the long edge of the sheet with the chalk line. Continue the row of sheets along the chalk line across the building.
  Important Note: *Do not* fasten down the entire first sheet before proceeding to the next; tack only the four corners (either screw down or leave heads of nails protruding) until the entire row of sheets is laid down. This will allow adjustments to be made if the sheets do not quite align.
- **Step 3**—Before fully fastening the sheets, lay out the OC spacing *along the edge of the sub-floor material*. This allows one to push bowed joists to align with the layout marks before fastening the sheets to the joists. Joists will often bow slightly left or right when framing; it is necessary to correct this, otherwise subsequent rows of sheets may not properly land on the joists.

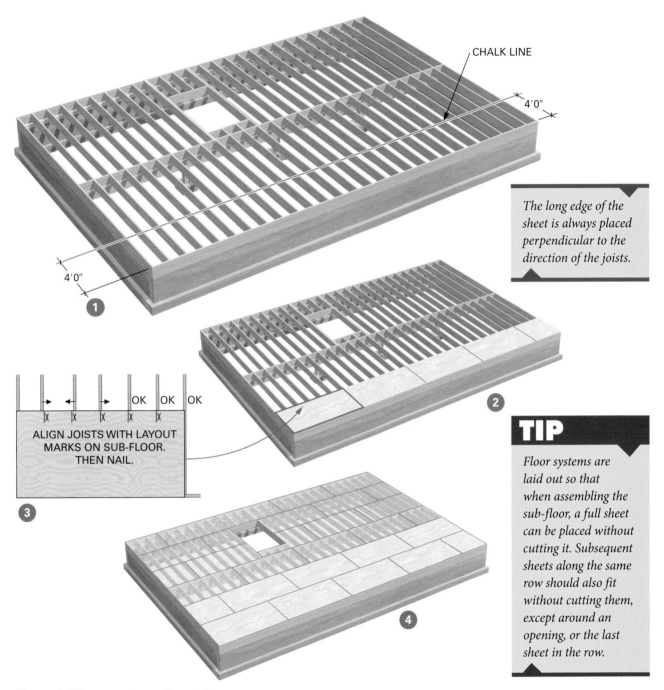

CHALK LINE

4'0"

4'0"

The long edge of the sheet is always placed perpendicular to the direction of the joists.

OK   OK   OK

ALIGN JOISTS WITH LAYOUT MARKS ON SUB-FLOOR. THEN NAIL.

TIP

Floor systems are laid out so that when assembling the sub-floor, a full sheet can be placed without cutting it. Subsequent sheets along the same row should also fit without cutting them, except around an opening, or the last sheet in the row.

Figure 4-37 Procedure for installing sub-floor

- **Step 4**—Start the second row with a half-sheet, or the cut-off from the first row, and continue with full sheets. Make sure to add construction adhesive before placing each row of the sub-floor material.

Fastening the Sub-Floor

Manufacturers of sub-floor material have recommendations regarding the placement of fasteners. Using too many fasteners will waste nails/screws and time. Not enough fasteners may cause a weakness and be a building code violation; furthermore, it may void a manufacturer's warranty. Often carpenters prefer to fasten the sub-floor to the joists with a combination of construction adhesive and screws. This is the best combination to prevent squeaky floors.

The Appendix 5 has a recommended nailing schedule for different types of materials.

> *Most codes require sub-floor fasteners to be located a minimum of 6" OC on the edges and 12" OC in the field.*

CANTILEVERED AREAS

Some houses may have a small section of floor that is cantilevered, such as the bay window area shown in Figure 4-38.

TIP

> *When framing a cantilevered floor section, reverse the crown of the joist so that the crown is facing downward. This will ensure that the cantilevered portion of the floor, as weight is added, will tend to level out.*

Figure 4-38 Allowable cantilever spans

- Cantilever backspan must be 3:1.
- Span of the cantilever cannot exceed the nominal depth of the wood floor joist.
- The maximum span of the cantilever is based on the table below.

Member Size and Spacing	Maximum Cantilever Span								
	Ground Snow Load								
	Up to 20 PSF			30 PSF			50 PSF		
	Roof Width			Roof Width			Roof Width		
	24 ft.	32 ft.	40 ft.	24 ft.	32 ft.	40 ft.	24 ft.	32 ft.	40 ft.
2 × 8 @ 12" OC	20"	15"	—	18"	—	—	—	—	—
2 × 10 @ 16" OC	29"	21"	16"	26"	18"	—	20"	—	—
2 × 10 @ 12" OC	36"	26"	20"	34"	22"	16"	26"	—	—
2 × 12 @ 16" OC	—	32"	25"	36"	29"	21"	29"	20"	—
2 × 12 @ 12" OC	—	42"	31"	—	37"	27"	36"	27"	17"
2 × 12 @ 8" OC	—	48"	45"	—	48"	38"	—	40"	26"

Figure 4-39 Cantilevered balcony spans

Back span example: a balcony with a 36" cantilever needs an additional 72" of length inside the structure.

- Cantilever backspan must be 2:1.
- Span of the cantilever cannot exceed the nominal depth of the wood floor joist.
- The maximum span of the exterior balcony cantilever is based on the table below.

Member Size and Spacing	Maximum Cantilever Span		
	Ground Snow Load		
	Less Than 30 PSF	50 PSF	70 PSF
2 × 8 @ 12" OC	42"	39"	34"
2 × 8 @ 16" OC	36"	34"	29"
2 × 10 @ 12" OC	61"	57"	49"
2 × 10 @ 16" OC	53"	49"	42"
2 × 10 @ 24" OC	43"	40"	34"
2 × 12 @ 16" OC	72"	67"	57"
2 × 12 @ 24" OC	58"	54"	47"

Other houses may have the entire length of a wall cantilevered over the foundation, such as the raised ranch in Chapter 1 (Figure 1-3). Regardless, special care must be taken when subjecting framing materials to forces of cantilevering. Guidelines to determine allowable cantilever are different, depending on the use. A load-bearing wall must have a cantilever back span (ratio of the joist that is inside the structure vs. the exterior cantilevered amount) of 3:1, while a balcony without a roof can have a cantilever back-span ratio of only 2:1. See tables for allowable cantilever spans. Furthermore, codes may be even more restrictive in areas that are earthquake prone or areas of high wind.

Because snow load weight transfers to the cantilevered portion of the floor system, stress loads often are individually engineered. Thus, the amount of cantilever allowed will vary regionally.

ENGINEERED LUMBER OPTIONS

Many types of engineered lumber are available today. Different types have their own specifications for installation, nailing schedules, bracing details, boring and notching instructions, cutting instructions, and even jobsite storage requirements. Many manufacturers specify their own hardware for installation. Some types of engineered lumber may have to be oriented correctly (the proper side facing up). Due to the various engineered products available, and the continuous changes, building code regulations often defer to manufacturer's recommendations. Generally, when purchasing an engineered lumber floor system, an informational packet will be available to the builder so that proper construction techniques are followed. Below is a brief overview of several popular residential options.

Truss Joists

Wood open-web truss joists can **free-span upward of 30′**. Their open-web design allows for mechanicals to be easily run without the time-consuming process of boring holes. Some materials are engineered to allow trimming a few inches. They are available in wood or a wood/steel combination.

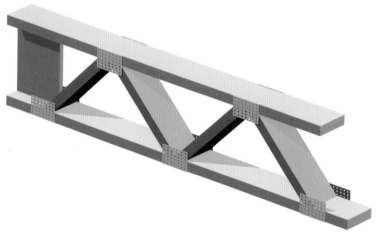

Figure 4-40 Truss-type joist

Wood I-Beam–Type Joists

These joists have an "I" shaped cross section, hence the name. They are available in different lengths of varying sizes and can be easily cut without compromising them. They can **free span more than 30′** and are available in much longer lengths, providing that intermediate support is utilized. Some manufacturers

TIP

Engineered lumber such as I-joists may have different specifications for temporary and permanent bracing.

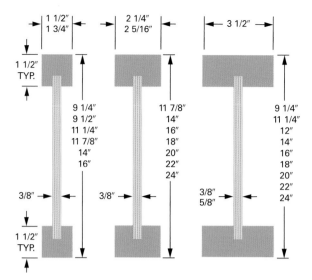

Figure 4-41 Cross-section of typical wood I-joist sizes

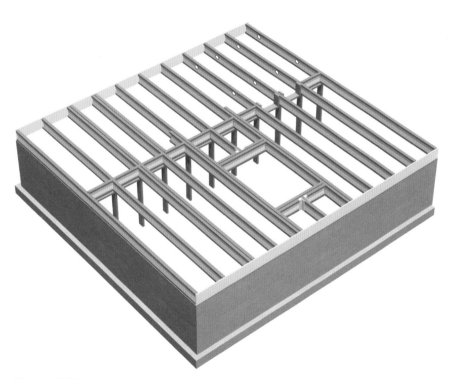

Figure 4-42 Engineered lumber floor system

incorporate perforated knock-out holes to speed up the process of installing mechanicals. They can also be used as rafters.

Engineered floor systems can be designed for virtually any configuration. Most manufacturers also offer complimentary hardware, such as joist hangers.

Laminated Veneer Lumber (LVL)

LVL beams are generally used where support for a heavy load is needed, such as a header or girder. Due to expense, they are not typically used for joists or rafters.

Figure 4-43 LVL beam

LVL beams are dense and heavy, they are essentially a beam made of plywood. Often two or more pieces are fastened together to allow them to free-span much longer distances than conventional solid lumber. Manufacturers will give special fastening requirements when building up two or more LVL beams. An example is shown in Figure 4-44, but make sure to check the manufacturer's recommendations.

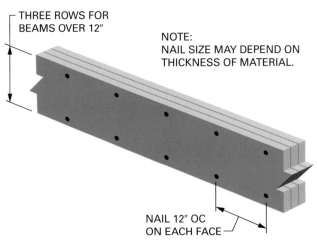

THREE ROWS FOR
BEAMS OVER 12"

NOTE:
NAIL SIZE MAY DEPEND ON
THICKNESS OF MATERIAL.

NAIL 12" OC
ON EACH FACE

Figure 4-44 Engineered lumber such as LVL beams may have specific nailing requirements

Laminated Strand Lumber (LSL) and Parallel Strand Lumber (PSL)

Of these two types, PSL (Figure 4-45) can be used where a stronger structural member is necessary, such as a girder or header. PSL beams are made of long thin strands of material laminated together.

Figure 4-45 PSL beam

LSL beams are made of smaller chips of wood glued together and can be used for shorter headers in light structural situations where strength is not as much of an issue (see Figure 4-46). They are sometimes used as rim joists when using an engineered lumber floor system.

Figure 4-46 LSL beam

Glued Laminated Timber (Glulam) Beams

Made by stack laminating solid lumber such as 2 × 6's, glulam beams are able to support heavy loads over long spans. The beams may have to be oriented in a specific way in order to perform correctly. For the longer beams, camber is built into them so that when weight is added they will settle flat. Many sizes are available and custom shapes such as arches can be manufactured.

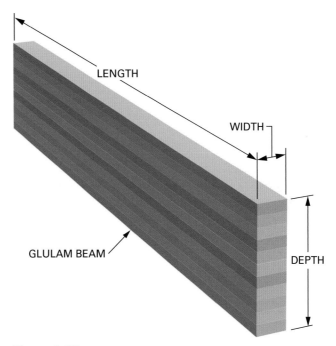

Figure 4-47 GLULAM beams are made of laminated solid lumber

Engineered Panels

Some types of engineered panels are unsuitable for sub-floor material. Different types have different uses. For sub-flooring, make sure to select sheets rated for the proper load and span. Select the appropriate thickness depending on the joist spacing. In the past, individual boards were used, and, more recently, plywood. Today, there are several alternative products, such as Oriented Strand Board (OSB). Like plywood, it also is available tongue and groove, which provides a tight fit and will not separate mid-span between joists.

The thickness of the material used sometimes depends on the OC spacing of the joists. In most situations, people use sub-floor material that is about 3/4″ thick. As always, follow building codes, blueprints, and manufacturers' recommendations for installation.

WALL FRAMING

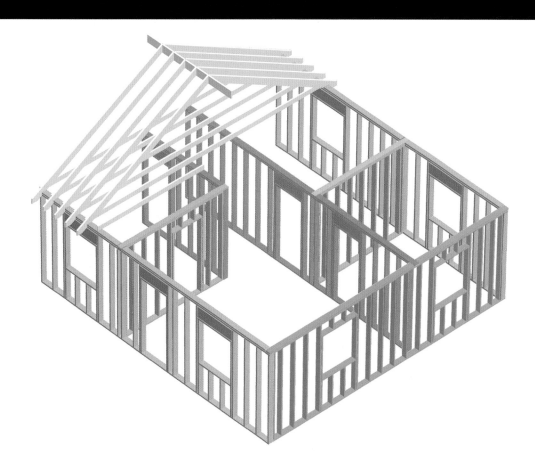

WALL FRAMING

Chapter Five Glossary

Terminology may vary regionally.

Refer to Figure 5-1 for the locations of the framing members described below.

Bottom plate—A horizontal framing member found at the base of a wall or partition. It is fastened to the bottom end of the studs.

Bracing—A support member that generally runs diagonally. Bracing can be temporary or permanent depending on the circumstances. There are different types of bracing.

Centerline—On plans there are measurements to the center of an opening or the center of a partition; these are centerline measurements—*not to be confused with OC.*

Cripple stud—A shortened stud generally found above a header or below a sill. Indicated with a "C" in layout.

Double top plate—Attached to the top of the top plate. It overlaps intersecting partitions and corners.

Header—Horizontal framing member that spans the top of an opening. It is strong enough to carry extra weight from above.

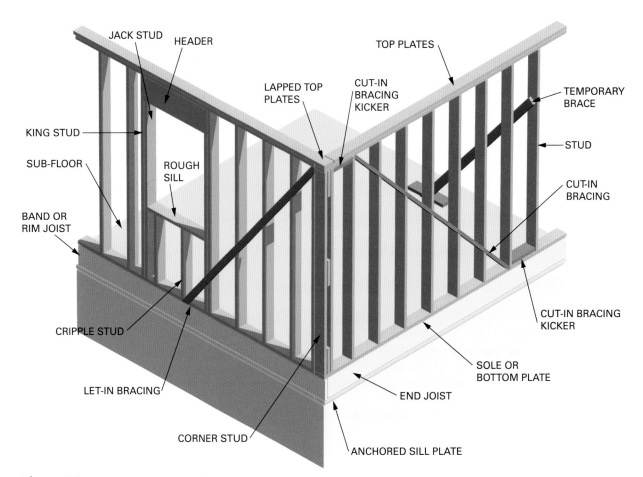

Figure 5-1 Framing components of a wall system

Jack stud—A shortened stud that carries a header. Indicated with a "J" during layout. Also called a trimmer stud.

King stud—A full-length stud, generally attached to a jack stud at a window or door opening. Indicated with an "X" during layout.

Load bearing—This type of wall or partition carries extra weight from above.

Non-load bearing—This type of a wall or partition does not carry any extra weight from above.

On center—Abbreviated "OC." Refers to the regular spacing of the framing members, the spacing from the center of one member to the center of the next—*not to be confused with centerline.*

Partition—Refers to an interior wall. Generally 2 × 4 unless mechanicals are planned to be hidden inside.

Plumb—Perfectly vertical.

Top plate—A horizontal framing member found at the top of a wall or partition. It is fastened to the tops of the studs.

Wall—Refers to an exterior wall. Can be either 2 × 6 or 2 × 4.

WALL/PARTITION LAYOUT

The basic concepts of wall/partition layout are the same as floor (box header) layout. For floor layout information, see Chapter 4.

Wall layout differs because there are four types of walls/partitions and each type is treated uniquely. The key is to correctly locate centerlines and the first stud (not counting the stud on the end of the plate), and then the remaining layout process is the same for all wall/partition types.

Identifying Wall/Partition Type

The first step of laying out a wall is to determine the wall type. In this book, the term "wall" will most often refer to perimeter (outer) walls of a structure. A "partition" will be considered an interior wall.

Note the double-ended arrow shown in Figure 5-2 with the word JOISTS above it; this indicates the direction that the floor joists are oriented. Generally, load-bearing walls/partitions run *perpendicular* to the direction of this arrow (which makes them perpendicular to the joists), and non–load-bearing walls/partitions run *parallel* to the joists. Also see Figure 5-5.

WALL TYPES

Load-Bearing Wall (LBW)

Often a LBW is the first wall to be framed. LBWs are generally oriented along the long axis of the house. Layout starts from the *same corner that the floor joist layout started*; this way, studs are located directly over the joists. This alignment from stud to joist is good for two reasons:

1. The load is transferred directly from the studs to the joists beneath.

2. The alignment makes it easier to run the mechanicals (pipes, wiring, and ductwork) between the floor joists and into the wall and partition cavities.

See Figure 5-6 for the LBW layout process.

NOTE

The first stud must be offset ½ the stud thickness so that the OC measurement is to the center of the stud. For example, when framing a load-bearing wall (LBW) with 16" OC spacing, the edge of the first stud is located 15¼" from the end of the plate, the second stud 16" from the first or 31¼" from the end of the plate, etc. See Figure 5-10.

Exterior walls are most often constructed of 2 × 6s or 2 × 4s. A 2 × 6 wall can carry more weight, but a 2 × 4 wall is generally adequate. The primary reason for using 2 × 6s for wall studs is to allow for thicker cavity insulation. In harsh northern climates, 2 × 6 walls are commonly used.

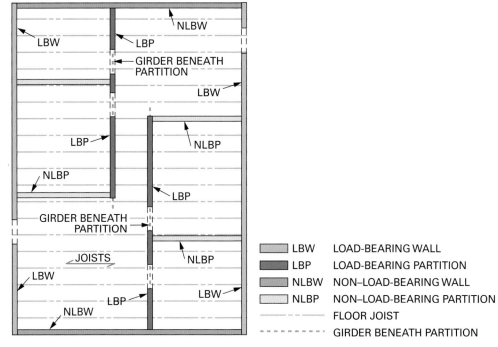

NOTE: IN THIS VIEW THE STARTING AND ENDING
LOCATIONS OF EACH OF THE WALL TYPES
DO NOT SHOW THE OVERLAPPING TOP PLATES.

Figure 5-2 There are four different wall/partition types in a typical structure

Non–Load-Bearing Wall (NLBW)

A NLBW abuts the LBW and is often built after the first load-bearing wall has been placed. There are two ways to frame the NLBW, and there is only a subtle difference between the methods.

- **Overlapping method**—One approach is to lay out the studs so when the sheathing of the NLBW is applied, it overlaps the LBW sheathing (see Figure 5-3).

NOTE

Overlapping (closed) corners are especially important when continuous insulated (CI) sheathing is used. CI is required in some areas to meet new energy codes. This overlap will prevent a thermal break at the corner.

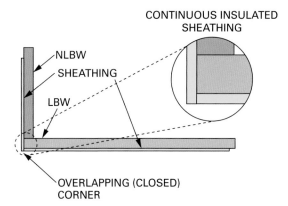

Figure 5-3 Overlapping method—the LBW is built and sheathed first, the NLBW abuts it, and the sheathing of the NLBW overlaps

- **Non-overlapping method**—The other method, seen in Figure 5-4, is to lay out the wall so the sheathing of the NLBW ends at the edge of the stud on the LBW (not overlapping the sheathing). This leaves a non-overlapping, open corner and is sometimes used when framing with non-insulated sheathing such as plywood or OSB (see Figure 5-4).

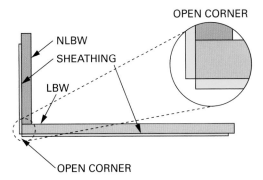

Figure 5-4 Non-overlapping method

See Figure 5-7 for the NLBW layout process.

Load-Bearing Partition (LBP)

This type of partition is often located near the center of the house, over a girder. It runs parallel to the LBWs (see Figure 5-2). It is important to lay out the studs to align with the joists beneath.

Non–Load-Bearing Partition (NLBP)

These partitions run in the same direction as the joists; therefore alignment with joists is not an issue (see Figure 5-2). OC layout can start at the end of the wall plate.

STEP-BY-STEP PROCESS FOR WALL LAYOUT

Start the layout from the same end of the building that the floor joist layout was performed, otherwise adjustments will be necessary to ensure that studs align with the joists below. The layout will be performed on either the top or bottom plate, or by placing both plates together, the layout marks can be simultaneously placed on both plates.

1. Determine wall type (LBW, NLBW, LBP, or NLBP).
2. Locate and mark openings, such as windows and doors. It is important to make the distinction between OC and centerline. Remember, centerline refers to the center of an opening or a partition. OC refers to the spacing of the framing members (studs in this case).
3. Locate and mark wall and partition intersections; Steps 2 and 3 can be reversed or performed simultaneously.
4. Locate the first and subsequent OC studs; there is always a stud at the end of the wall; do not count this one as the first stud.

Some carpenters, out of habit, overlap all corners. While it may not be necessary, it is considered good technique. One drawback with this method is that if the structure width is a multiple of 4 feet, full sheets of sheathing will just cover the framing; if overlapping sheathing at the corners, a narrow strip of sheathing will be needed to complete the job.

Interior walls (called partitions) are nearly always 2 × 4 unless otherwise indicated. Occasionally it is necessary to build a 2 × 6 partition in order to house plumbing or other mechanicals within the stud cavity.

TIP

It is good practice to frame any partition running parallel to an LBW so that the studs align with joists below; this will make installation of mechanicals easier.

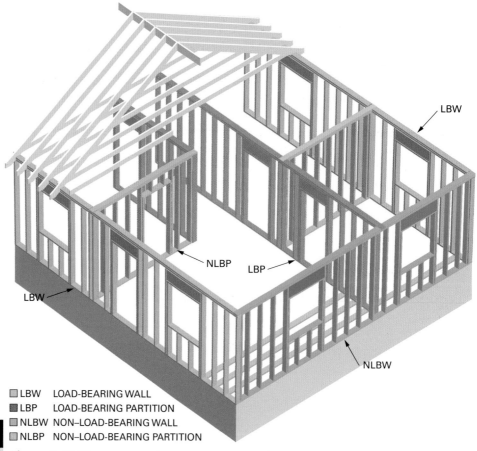

LBW   LOAD-BEARING WALL
LBP   LOAD-BEARING PARTITION
NLBW   NON–LOAD-BEARING WALL
NLBP   NON–LOAD-BEARING PARTITION

Figure 5-5 Walls are exterior and partitions are interior. Either may be load bearing or non-load bearing

Load-Bearing Wall (LBW)

Lay out a load-bearing wall by measuring OC studs and centerlines from the *end* of the wall plate (Figure 5-6).

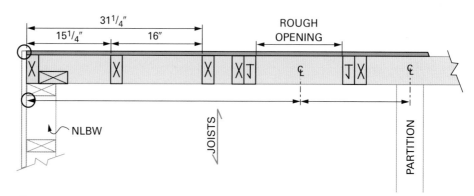

Figure 5-6 Load-bearing wall layout

Non–Load-Bearing Wall (NLBW)

The method shown in Figure 5-7 ensures *overlapping* of the sheathing as shown in Figure 5-3. Lay out OC studs by including the LBW wall thickness *and* sheathing thickness. Lay out the centerlines by including *only* the wall thickness (not the sheathing). One way to perform the NLBW OC layout is to let the end of tape measure overhang the end of the plate the thickness of the LBW plus the sheathing thickness. For example, if abutting a 2 × 6 LBW with 1″ insulated sheathing, the end of the tape measure will hang over the end of the NLBW plate 6½″. From there, the first OC stud will be marked at 15¼″. This first stud will actually be located 8¾″ from the end of the wall plate. The edge of the second stud will be located 16″ from the edge of the first. To locate centerlines, allow the tape measure overhang the plate only 5½″.

> *Non-overlapping method: In order to lay out the OC wall studs so the sheathing does not overlap, only include the LBW wall thickness,* not the *sheathing thickness, for centerlines and OC studs.*

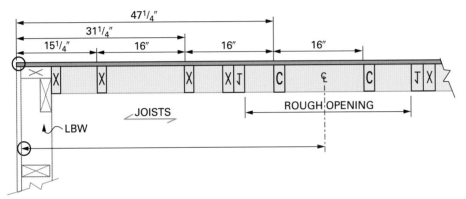

Figure 5-7 Non–load-bearing wall layout

Load-Bearing Partition (LBP)

Lay out OC studs *and centerlines* (from the building line) by including the adjoining wall thickness; this will ensure the studs are placed over the floor joists. See Figure 5-8. Let the end of the tape measure hang over the end of the LBP plate, the thickness of the NLBW. For example, if abutting a 2 × 6 exterior wall, let the tape

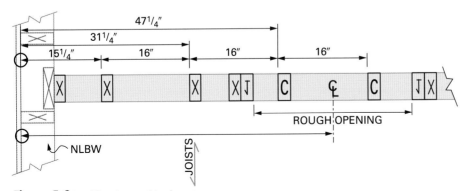

Figure 5-8 Load-bearing partition layout

measure hang past the end of the plate 5½″ to mark the first OC stud and when marking the centerlines.

Non–Load-Bearing Partition (NLBP)

> *A NLBW OC stud layout can begin at the end of the plate because the joists below are running parallel to the direction of the wall; therefore, alignment with joists is not an issue.*

Lay out the centerlines (from the building line) by including the thickness of the adjoining wall. See Figure 5-9. Lay out OC studs from the end of the plate. For example, when NLBP is abutting a 2 × 6 wall, allow the tape measure to overhang the plate by 5½″ to mark the centerlines. To locate the OC studs, simply measure from the end of the plate.

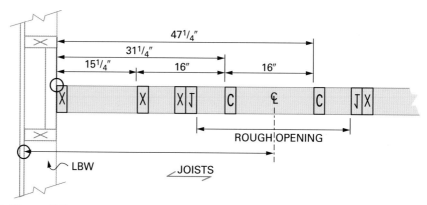

Figure 5-9 Non–load-bearing partition layout

Layout of an Entire Wall

> *19.2″ OC layout, although not often used, is a valid OC spacing option. Take the time to look at a tape measure and notice the small red or black diamonds located every 19.2″— most 25′ or longer tape measures have 19.2 OC notations to aid layout. One drawback is that fiberglass batt insulation in this width is not readily available.*

Typically, layout of both the top plate and bottom plates is done simultaneously. A square will allow the transfer of marks from one plate to the other. This ensures that when the wall is assembled, the spacing will be identical on the top and bottom plates. Figure 5-10 shows an overview of the process of laying out a load-bearing wall.

Notes on Stud Spacing

Carpenters often have the habit of framing 16 OC, when 24 OC may be adequate. Furthermore, because of an effect called thermal bridging, a 24 OC wall is slightly more energy efficient than one framed 16 OC. Wood studs do not insulate as well as fiberglass and other types of insulation. They conduct heat (cold) more quickly than insulation; therefore, fewer studs in an insulated wall will translate to slightly better energy efficiency. There is also the added bonus of using less framing material! Codes specify when 24 OC framing is adequate. See table in Chapter 5 Appendix. Some consider 24″ OC framing synonymous with a concept called advanced framing. For more information on advanced framing details, see the Advanced Framing section later in this chapter.

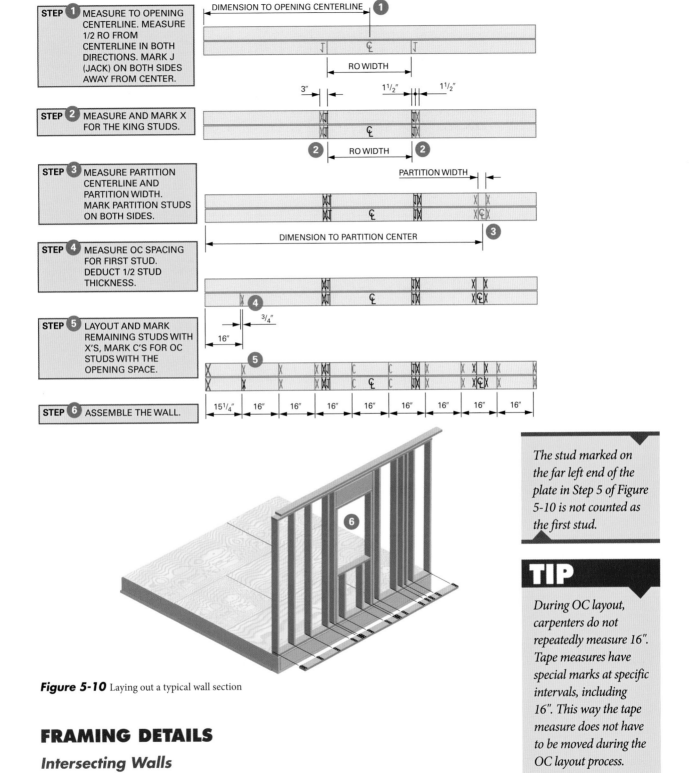

STEP 1 MEASURE TO OPENING CENTERLINE. MEASURE 1/2 RO FROM CENTERLINE IN BOTH DIRECTIONS. MARK J (JACK) ON BOTH SIDES AWAY FROM CENTER.

STEP 2 MEASURE AND MARK X FOR THE KING STUDS.

STEP 3 MEASURE PARTITION CENTERLINE AND PARTITION WIDTH. MARK PARTITION STUDS ON BOTH SIDES.

STEP 4 MEASURE OC SPACING FOR FIRST STUD. DEDUCT 1/2 STUD THICKNESS.

STEP 5 LAYOUT AND MARK REMAINING STUDS WITH X'S, MARK C'S FOR OC STUDS WITH THE OPENING SPACE.

STEP 6 ASSEMBLE THE WALL.

Figure 5-10 Laying out a typical wall section

The stud marked on the far left end of the plate in Step 5 of Figure 5-10 is not counted as the first stud.

TIP

During OC layout, carpenters do not repeatedly measure 16". Tape measures have special marks at specific intervals, including 16". This way the tape measure does not have to be moved during the OC layout process.

FRAMING DETAILS

Intersecting Walls

Figure 5-11 shows an LBW and an NLBW abutting to form a corner of the building.

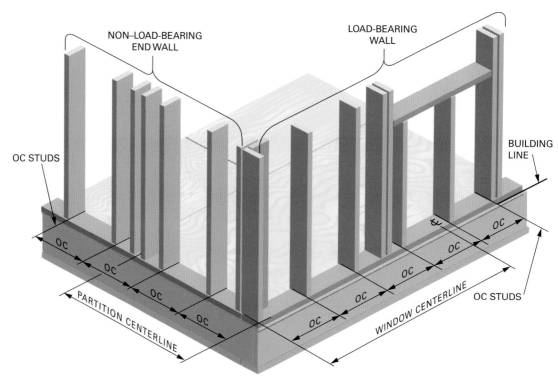

Figure 5-11 Layout wall components by measuring from the building line

Corner Details

When two walls meet at the corner, there are different ways to frame this detail. Figures 5-12 and 5-13 shows two common methods. The corner in the diagram on the left will have to be pre-insulated before sheathing is applied, whereas the corner on the right can be insulated after sheathing has been completed.

> *See Advanced Framing section for a two-stud corner.*

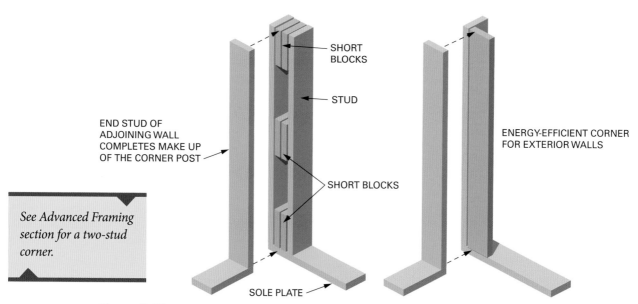

Figure 5-12 Construction of corner posts

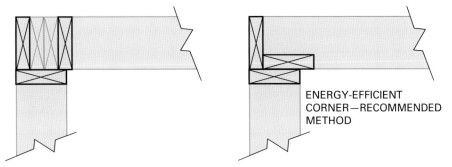

EITHER OF THESE CORNERS CAN BE BUILT WITH 2 × 4 OR 2 × 6 FRAMING; THERE MAY BE A SLIGHT VARIATION. STUDS SHOWN REPRESENT 2 × 6s.

ENERGY-EFFICIENT CORNER—RECOMMENDED METHOD

Figure 5-13 Layout notations (for corners) on the plates

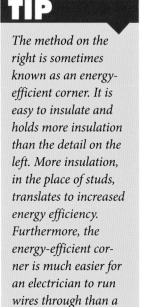

TIP

The method on the right is sometimes known as an energy-efficient corner. It is easy to insulate and holds more insulation than the detail on the left. More insulation, in the place of studs, translates to increased energy efficiency. Furthermore, the energy-efficient corner is much easier for an electrician to run wires through than a closed or solid corner.

Intersecting Partitions

The diagram of the corner on the left of Figure 5-14 illustrates a closed corner that needs to be pre-insulated (before sheathing); therefore it is best used only for partition-to-partition intersections, not for a partition to an exterior wall intersection. The other two corners (middle and right) can be insulated after framing and sheathing are complete. These methods are particularly useful when remodeling and wiring is already in place. See Figure 5-15 for another view.

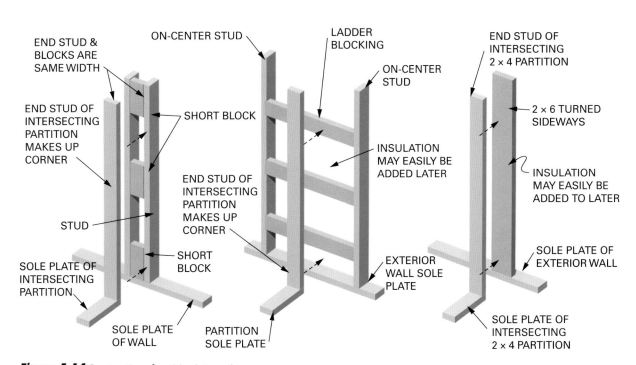

Figure 5-14 Construction of partition intersections

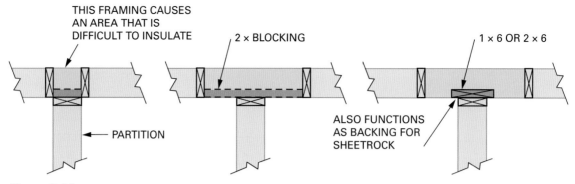

THIS FRAMING CAUSES
AN AREA THAT IS
DIFFICULT TO INSULATE

2 × BLOCKING

1 × 6 OR 2 × 6

PARTITION

ALSO FUNCTIONS
AS BACKING FOR
SHEETROCK

Figure 5-15 Layout marks on plates correspond to isometric drawings shown in Figure 5-14

Wall Plates

Wall plate material should be the straightest material, as this will make the construction process easier and more accurate. When building long walls where more than one plate is required, joints between plates should be located at the center of a stud (see Figure 5-16).

When plates intersect, the doubled top plates should overlap whenever possible (see Figure 5-17). This overlapping ties the walls together and adds strength to the structure.

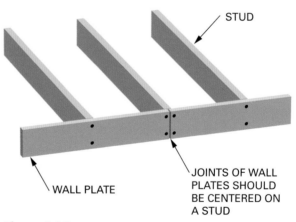

STUD

WALL PLATE

JOINTS OF WALL
PLATES SHOULD
BE CENTERED ON
A STUD

Figure 5-16 Joints in the plates should fall at the center of the stud

TIP

Some carpenters frame with single top plates in order to reduce material and thermal bridging effects, but metal plates must span the joints and intersections. Single plates may not be acceptable for some building codes. See Advanced Framing section later in this chapter for more details.

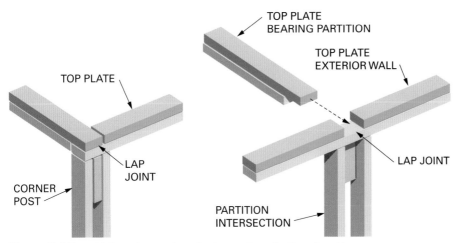

Figure 5-17 Doubled top plates are lapped at intersections of walls and partitions

Headers

Headers span openings and are sized to carry weight. Occasionally carpenters choose to use larger headers than are required. This can save time by avoiding the task of cutting and placing several cripple studs. This will, however, reduce the amount of insulation that can be used. Figure 5-18 shows a framed window opening with cripple studs above the header. Figure 5-19 illustrates a door opening framed with a larger header and no cripple studs.

Codes generally allow for a small opening in an NLBP, or NLBW, to be framed without the use of a structural header, as shown in Figures 5-20 and 5-21.

There are many types of headers (see Figures 5-22 a & b); the span, load, and header material type will determine the necessary size. See Appendix 3 for header span charts. Always consult a design professional.

Typically headers less than 6' wide are supported by a single jack stud on each side (see Figure 5-18); this makes the header 3" longer than the rough opening (RO) width. For more information see Appendix 3.

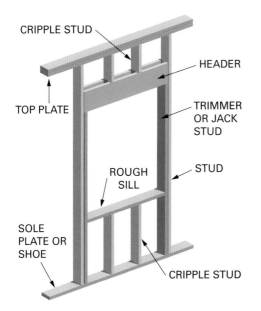

Figure 5-18 Typical framing for a window opening

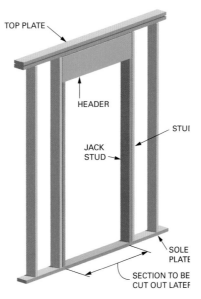

Figure 5-19 Typical framing for a door opening

TIP

There is available a piece of hardware similar to a joist hanger that can be used to support a header. The hanger attaches to the king stud and supports the header, eliminating the need for a jack stud. This method reduces framing material, thus reduces thermal bridging. However, removing jack studs may cause other problems for later steps in the construction process, such as fastening windows, doors, siding, trim, electrical boxes, etc.

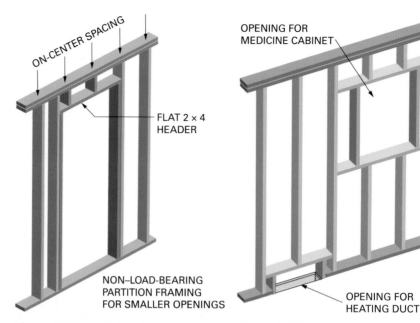

Figure 5-20 A method for framing a non–load-bearing header

Figure 5-21 Miscellaneous openings in interior partitions are framed with non–load-bearing headers

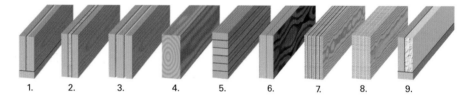

1. A BUILT-UP HEADER WITH A 2 × 4 OR 2 × 6 LAID FLAT ON THE BOTTOM.
2. A BUILT-UP HEADER WITH A $\frac{1}{2}''$ SPACER SANDWICHED IN BETWEEN.
3. A BUILT-UP HEADER FOR A 6″ WALL.
4. A HEADER OF SOLID SAWN LUMBER.
5. GLULAM BEAMS ARE OFTEN USED FOR HEADERS.
6. A BUILT-UP HEADER OF LAMINATED VENEER LUMBER.
7. PARALLEL STRAND LUMBER MAKES EXCELLENT HEADERS.
8. LAMINATED STRAND LUMBER IS USED FOR LIGHT-DUTY HEADERS.
9. ENERGY-EFFICIENT HEADER WITH RIDGED FOAM INSULATION.

Figure 5-22a Some of the different types and styles of headers

Placement of Headers

The height of a door header generally determines the height of a window header; this allows the outside appearance of a structure some uniformity. There are exceptions, however, such as high-gable end walls, specially shaped windows such as transoms, arches, and so forth. Typical residential doors are 6′ 8″ tall and vary in width from 24″ (small closet) upward to 36″ (or larger) for an entry door. Many carpenters choose to save time by installing pre-hung door units. These are already hung on hinges from the jambs and the exterior door units already have the thresholds applied. Manufacturers supply recommended rough openings (RO) for doors; however, many carpenters simply add 2½″ to the height of a door and 2″ or 2½″ to the width of the door. These dimensions will work as

TIP

Note Figure 5-22a #9—this is a popular method to build a header and is more energy efficient than a solid wood header. Figure 5-22b illustrates a detailed view.

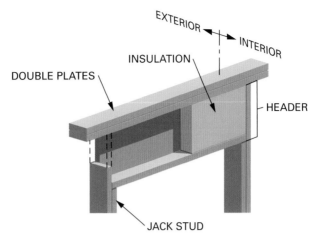

Figure 5-22b A cross section of an insulated header as it appears in a wall

RO dimensions for most standard pre-hung doors. For example, when framing an RO for a 3-0, 6-8 door (3′ 0″ wide and 6′ 8″ tall), add 2½″ to both the height and the width to come up with the RO, 3′ 0″ + 2½″ = 3′ 2½″ RO width. 6′ 8″ + 2½″ = 6′ 10½″ RO height. Make sure to cut the header 3″ wider than the RO width (3′ 2½″ + 3″) to allow it to rest on the jack studs, making the header 3′ 5½″.

Jack Stud Length

Pre-hung exterior doors have a threshold, side jambs, and a head jamb already attached; this will account for some of the extra 2½″ in height and width of the RO; the remaining space is to adjust (plumb and shim) the door into place before fastening.

If the RO door height is 6′ 10½″ (82½″), the jack stud length will be 81″ (82½″– 1½″) (bottom plate thickness) = 81″. Figures 5-23a and 5-23b show the relative elevations of the framing members.

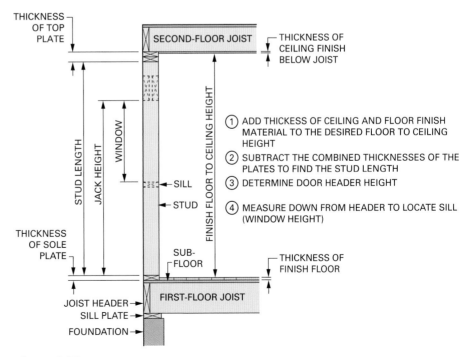

Figure 5-23a Vertical layout of wall framing components

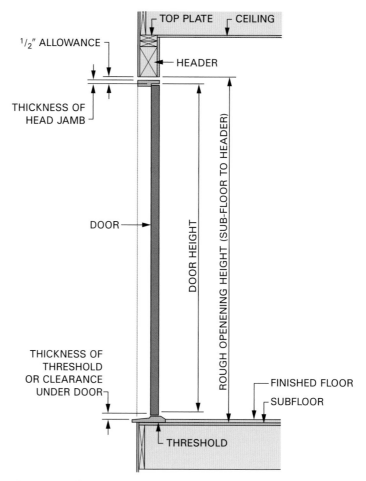

Figure 5-23b Determining the rough opening height of a door opening

When framing a window RO (height and width), there are so many manufacturers and different window sizes that windows are always supplied with an RO. Figure 5-24 shows a sample page from a manufacturer's catalog. Remember, window headers are generally placed at the same height as door headers. Thus, to locate the window sill, simply measure down from the bottom of the header the window's RO height. This will locate the top of the sill. Refer to Figure 5-23a above.

Story Pole

Some carpenters complete a vertical layout on a spare stud and use this to aid in repetitive vertical layout. Header height is represented on the story pole as well as the different sill heights of the windows. Anyone needing reference for layout purposes can use the story pole.

TIP

Using a story pole can be a time saver. Wall layout is only one of many ways a story pole can be utilized. They can be used on complex stairs, siding, and more.

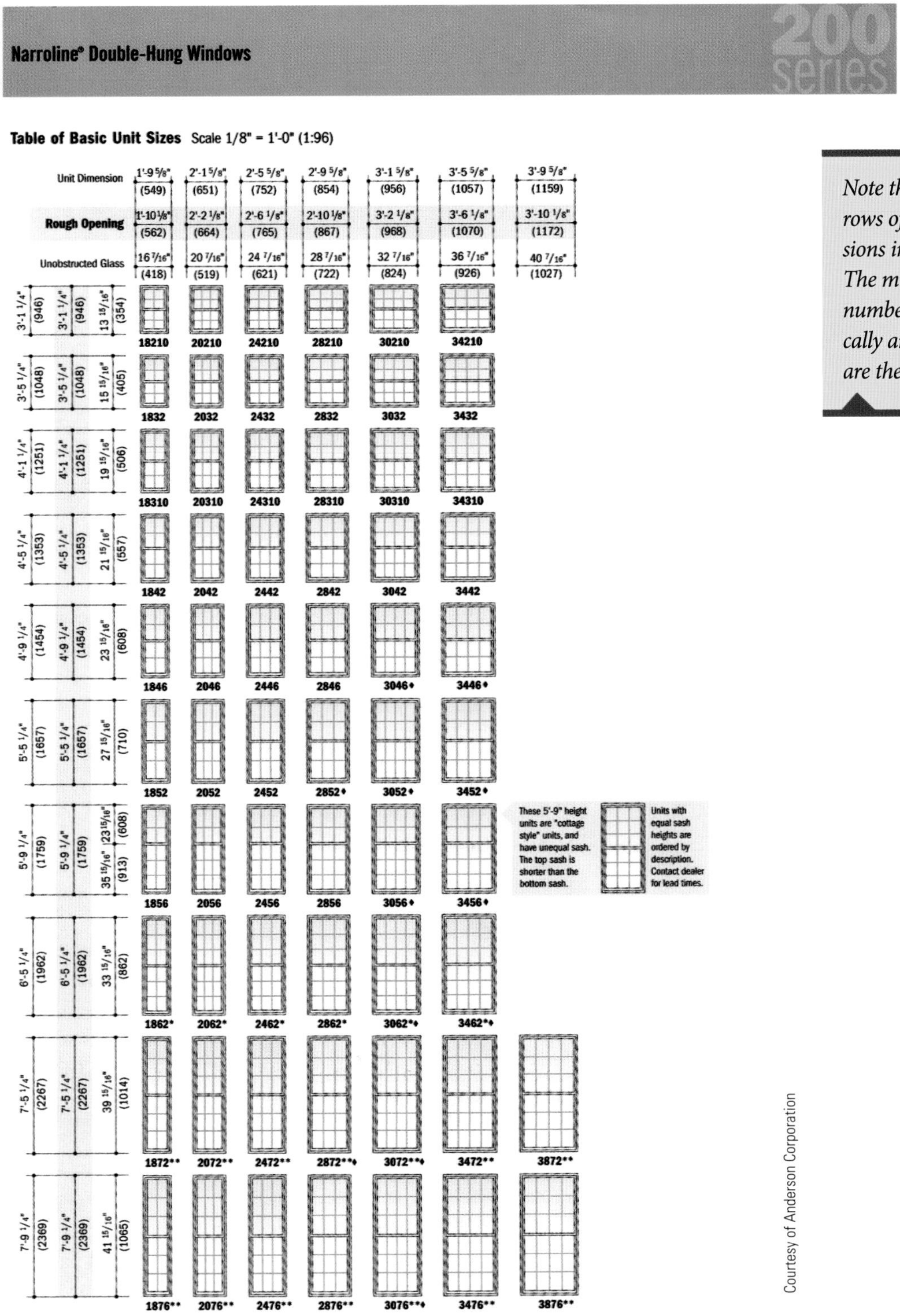

Narroline® Double-Hung Windows

Note the three rows of dimensions in Figure 5-24. The middle sets of numbers, both vertically and horizontally, are the RO dimensions.

Figure 5-24 Sample of a manufacturer's catalog showing rough opening sizes for window units

Placement of Nailers and Blocking

Nailers and blocking (backing) provide a means to attach sheetrock, cabinets, handrails, lighting, plumbing fixtures, and more. Strategic placement of blocking can help an electrician, a plumber, a trim carpenter, or a cabinet installer later on in the job. Figures 5-25 through 5-29 suggest locations for these nailers.

Some backing may be required in the ceiling to aid in securing parallel partitions (Figure 5-28).

Often scraps of 2× stud material are nailed on the king stud at the edge of a door opening in order to provide a surface for an electrical switch box to be placed. Spacing switches away from the doors is important so that the electrical box will not interfere with the door trim.

Often (scrap) pieces of 1× or 2× material are added to the tops of the doubled plates that parallel the ceiling joists. See Figure 5-29. This allows for backing to attach the ceiling.

A row of solid horizontal blocking at mid-span is generally required inside walls when the wall cavity is taller than 8′. This is to satisfy fire code requirements. Check local requirements!

Transitioning to Another Floor Level

When building a two-story structure, the second floor is typically framed on top of the doubled plate. Exterior walls of the structure receive another box header framed on top of them. This is similar to the way the box header is placed on the sill plate. Figures 5-30 and 5-37 illustrate this concept.

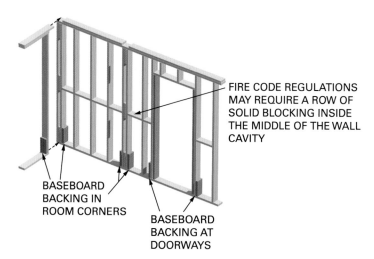

FIRE CODE REGULATIONS MAY REQUIRE A ROW OF SOLID BLOCKING INSIDE THE MIDDLE OF THE WALL CAVITY

BASEBOARD BACKING IN ROOM CORNERS

BASEBOARD BACKING AT DOORWAYS

Figure 5-25 Backing is sometimes installed for baseboard at corners and door openings and at mid-span for fire code regulations

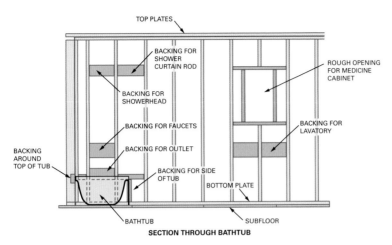

SECTION THROUGH BATHTUB

Figure 5-26 Typical backing needed in bathrooms

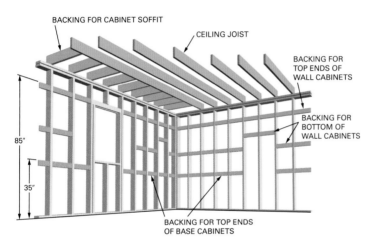

Figure 5-27 Considerable backing is needed in kitchens

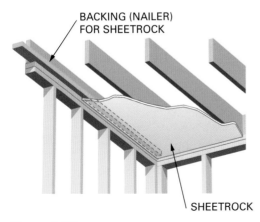

Figure 5-29 Backing is necessary on the tops of walls/partitions

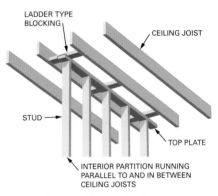

Figure 5-28 Ladder-type blocking provides support for the top plates of interior partitions that run parallel to joists

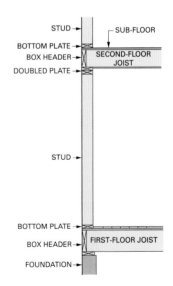

Figure 5-30 Framing the transition to the next floor

Load-bearing partitions will have the joists resting directly on them. Preferably the joists will be directly above the studs, as in Figure 5-31. Figure 5-32 shows an alternative to offsetting and overlapping the joists.

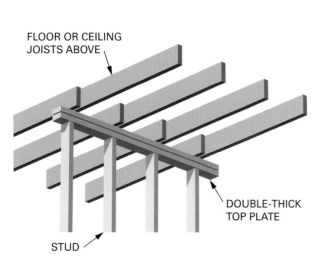

Figure 5-31 Load-bearing partitions support the weight of the floor or ceiling above

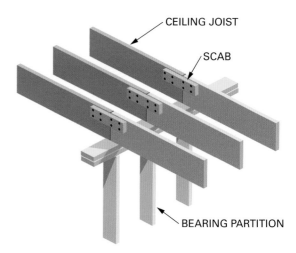

Figure 5-32 The joints of in-line ceiling joists must be connected with scabs (small connecting blocks of wood) at the bearing position

In the interior of a house there are often archways (not necessarily a rounded arch) between rooms. This generally indicates a header/beam that has been covered over. Archways are often framed to match door height. Figure 5-33 shows how this type of arch may be framed. In this example, one end of the header is supported by studs in the exterior wall.

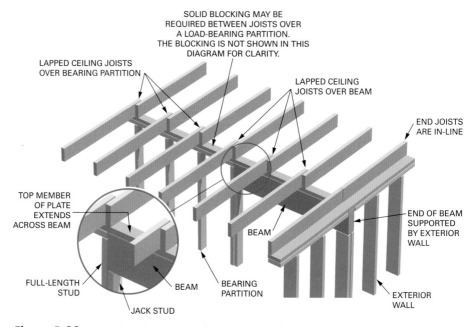

Figure 5-33 Support for ceiling joists may be placed as a header in the wall below

If there is an area of the house where an arch will detract from the appearance, the ceiling can be framed "flush" from one room to the next. See Figure 5-34. In this case, the header is framed flush with the ceiling joists, and the joists are secured to the header with joist hangers.

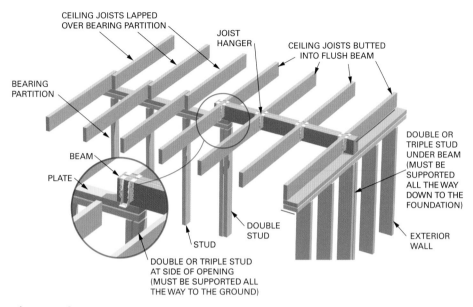

Figure 5-34 Support for ceiling joists may be a flush framed girder. This creates a flush ceiling through the opening below

Non–load-bearing partitions can be placed wherever needed, then fastened to appropriate blocking. An NLBP will not carry significant weight; however, joists below should be reinforced (see Figure 4-19).

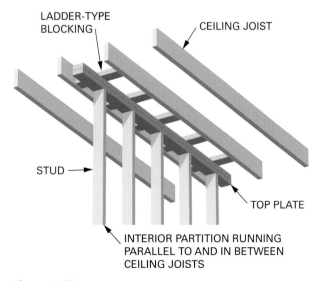

Figure 5-35 Non–load-bearing partitions can be placed anywhere

Ceiling Joists

In Figures 5-33 through 5-35, the second floor's floor joists also act as the first floor's ceiling joists. In addition to providing a place to attach the finished ceiling, ceiling joists have an important structural role as well. They tie together the outer walls of the structure and prevent the outside walls from bowing. Figure 5-36 illustrates this concept of forces.

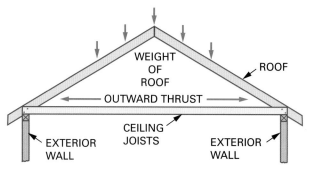

Figure 5-36 Ceiling joists tie the roof frame together into a triangle, which resists the outward thrust caused by the rafters

Structures can be built without the use of ceiling joists, but that requires some special details that will be covered in the next chapter (see "Cathedral Ceilings" section).

Framing Types

- **Platform frame construction**—Figure 5-37 illustrates the framing members of a two-story house. This framing method is commonly used today. At each floor level, a platform (floor) is built before continuing to the next level. This chapter had dealt with platform framing almost exclusively.

- **Balloon frame construction**—Very common in the past, but not used as much today see Figure 5-38. Studs are long and run from the sill plate to the top plate of the second story. Firestop blocking is required between floors. The Figure 5-39 detail illustrates how the second floor's joists are secured to the balloon frame.

- **Post-and-beam construction**—There are many variations of this type of construction. Typically larger framing members are used that are spaced farther apart. For example, in both variations of Figure 5-40, the roof planks will need to be 2″–3″ thick depending on the spacing between rafters, and the vertical posts will need to be 4 × 4s or 6 × 6s.

- **Advanced framing**—This is a variation of platform frame construction and was mentioned earlier in the chapter. This type of framing focuses on optimizing construction efficiency, without the use of any unnecessary material. The first diagram is Figure 5-41.

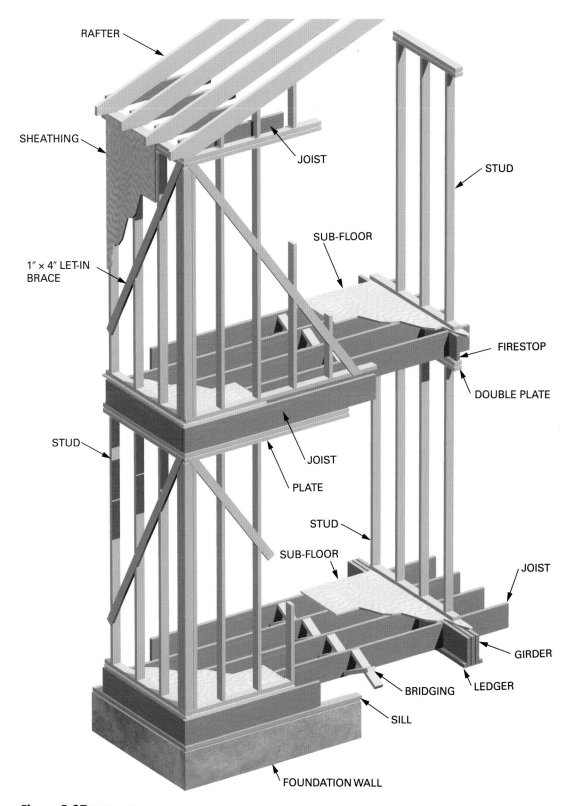

Figure 5-37 Platform frame construction

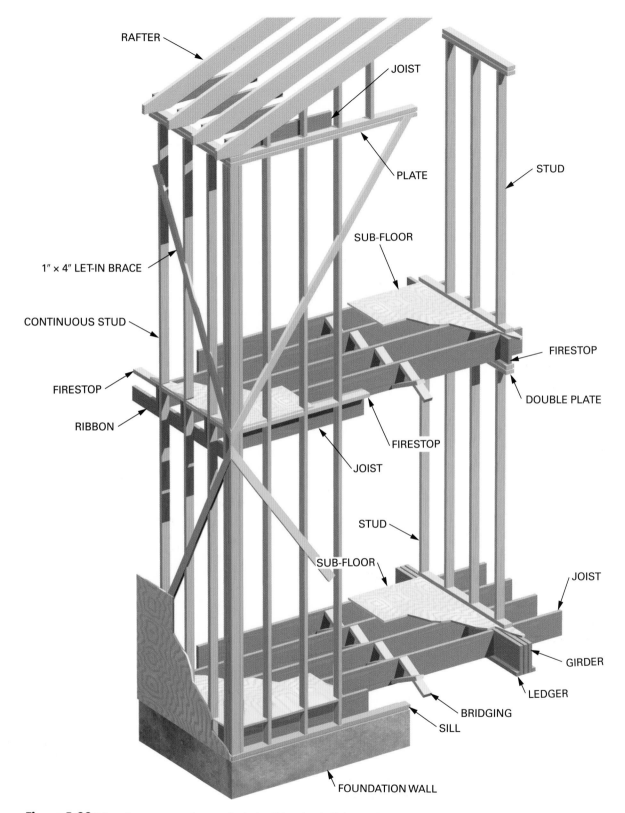

Figure 5-38 Balloon frame construction, popular in the 19th and early 20th centuries

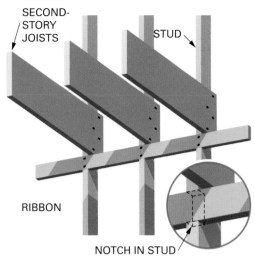

SECOND-STORY JOISTS

STUD

RIBBON

NOTCH IN STUD

Figure 5-39 Ribbons are used to support floor joists in balloon frame walls

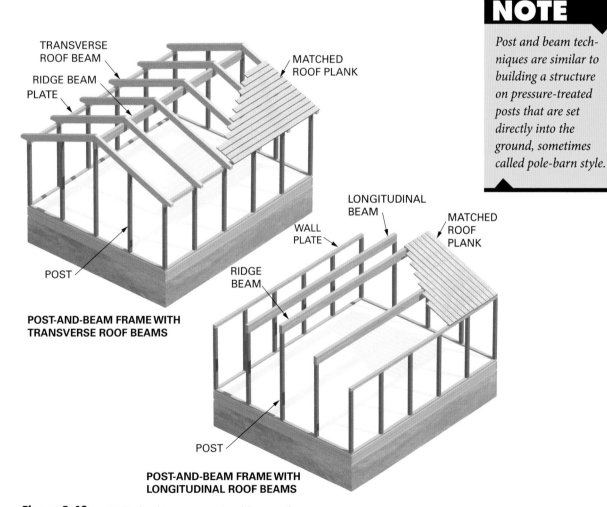

NOTE

Post and beam techniques are similar to building a structure on pressure-treated posts that are set directly into the ground, sometimes called pole-barn style.

TRANSVERSE ROOF BEAM

MATCHED ROOF PLANK

RIDGE BEAM

PLATE

POST

POST-AND-BEAM FRAME WITH TRANSVERSE ROOF BEAMS

LONGITUDINAL BEAM

MATCHED ROOF PLANK

WALL PLATE

RIDGE BEAM

POST

POST-AND-BEAM FRAME WITH LONGITUDINAL ROOF BEAMS

Figure 5-40 Longitudinal and transverse post-and-beam roofs

Advanced Framing

"Advanced framing" is a term that carpenters sometimes use to describe this method of building. Another term, probably more descriptive, is "value engineering," which has to do with producing a good product, a home in this case, without excessive expense or a loss in quality. As noted earlier in this chapter, this type of framing focuses on optimizing the structure without the use of unnecessary materials.

Too often, a construction method is chosen based on tradition or familiarity, instead of what is needed to produce a structurally sound building. As it turns out, many buildings are overbuilt and much more material used than necessary. This can increase the overall cost without increasing the overall value. Carpenters tend to build the way they were taught, and old habits are hard to break. Many people are comfortable adopting some of the below methods but not others.

In recent years, more and more attention is being paid to phrases such as "green building" or "sustainable building techniques." While these definitions include the environmental impact of a product, or the long-term sustainability of the structure, value engineering is also part of the equation.

Table 5-1 shows a comparison of some of the main differences of the two construction techniques. Following the table will be some explanations and diagrams.

Table 5-1 Comparing traditional framing with advanced framing

Traditional Framing	Advanced Framing
16″ OC framing	24″ OC framing
Doubled top plates	Single top plates
King stud and jack stud combination	King stud with header hanger
Built-up header	Single header
Three- or four-stud corner	Two-stud corner with F clips
Fewer design restrictions	Design & build with 2′ modules
Fewer construction restrictions	May need to "stack" the framing

The following bullet points show advantages/disadvantages of using advanced framing techniques.

16″ OC framing vs. 24″ OC framing (traditional vs. advanced framing). (See Table 5-1).

- When framing 24″ OC, for every 4 feet of framed wall, a stud is eliminated; this will save material (see Figure 5-41). This includes the elimination of any unnecessary cripple studs.
- When eliminating studs, some fasteners and labor to apply the fasteners will be eliminated.
- Thermal bridging will be minimized with 24″ OC framing (insulation insulates better than studs).

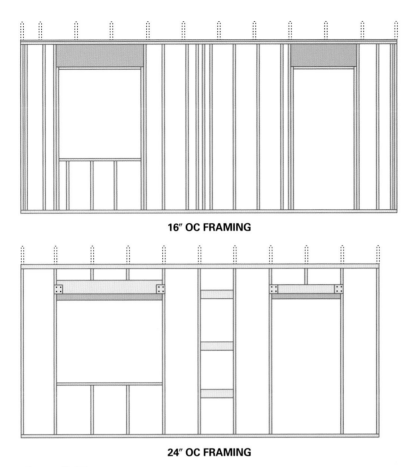

16″ OC FRAMING

24″ OC FRAMING

Figure 5-41 16″ OC vs. 24″ OC. 16″ OC framing (top diagram) uses much more material than 24″ OC framing, which is often more than adequate for the strength of a structure

- When framing 24″ OC is used, less drilling is required through studs/joists to place mechanicals.
- Codes allow a one-story house with 2 × 4s 24″ OC. 2 × 6s must be used if a two-story house with 24″ OC framing is desired.
- Extra blocking/backing will be necessary due to there being less framing material to fasten components to.
- Must make sure sheathing/siding and drywall is rated for 24″ OC framing.

Doubled top plates vs. a single top plate (traditional vs. advanced framing). (See Table 5-1).

- When framing with a single top plate, this will save plate material for the length of every wall on each floor of the house (see Figure 5-42).
- Some carpenters purchase pre-cut studs. They are already a specific length. It may be difficult or more expensive is some areas to get the studs 1½″ longer to make up for the second plate, or carpenters will have to take the time to cut all studs to length.

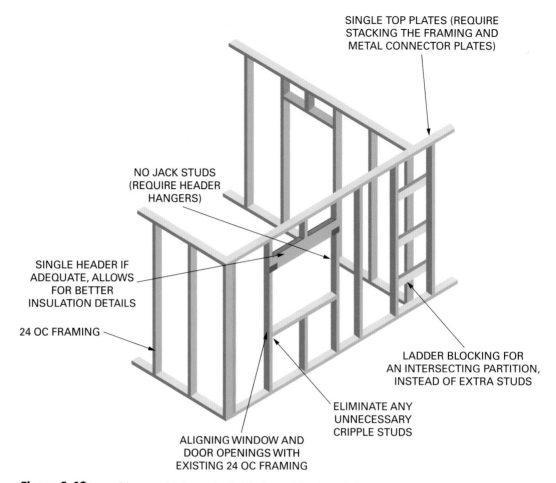

SINGLE TOP PLATES (REQUIRE STACKING THE FRAMING AND METAL CONNECTOR PLATES)

NO JACK STUDS (REQUIRE HEADER HANGERS)

SINGLE HEADER IF ADEQUATE, ALLOWS FOR BETTER INSULATION DETAILS

24 OC FRAMING

LADDER BLOCKING FOR AN INTERSECTING PARTITION, INSTEAD OF EXTRA STUDS

ELIMINATE ANY UNNECESSARY CRIPPLE STUDS

ALIGNING WINDOW AND DOOR OPENINGS WITH EXISTING 24 OC FRAMING

Figure 5-42 Some of the many details associated with advanced framing techniques

- If pre-cut batt insulation is used in combination with a single top plate, it will be too short.
- Single plates must be fastened together. Typically carpenters use metal connector plates attached to the top plates at wall/partition intersections (see Figure 5-43).
- If a two-story house is desired, not only will the studs have to be 2 × 6s, but the framing will have to be "stacked." This means that studs from one floor must be in line with other studs, joists, and rafters/trusses (see Figure 5-46).

King stud and jack stud combination vs. a king stud with a header hanger (traditional vs. advanced framing). (See Table 5-1).

- It eliminates two studs per door/window and will reduce thermal bridging.
- It requires additional hardware called a header hanger (see Figure 5-44).
- If only a king stud is used, it may make it difficult to fasten some types of trim.
- Some building codes, especially in high wind or seismic areas, may not allow the elimination of the jack studs.

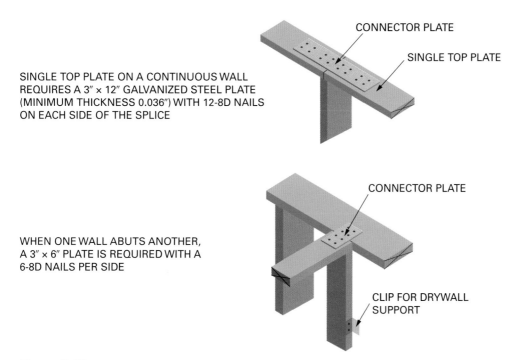

CONNECTOR PLATE

SINGLE TOP PLATE

SINGLE TOP PLATE ON A CONTINUOUS WALL
REQUIRES A 3″ × 12″ GALVANIZED STEEL PLATE
(MINIMUM THICKNESS 0.036″) WITH 12-8D NAILS
ON EACH SIDE OF THE SPLICE

CONNECTOR PLATE

WHEN ONE WALL ABUTS ANOTHER,
A 3″ × 6″ PLATE IS REQUIRED WITH A
6-8D NAILS PER SIDE

CLIP FOR DRYWALL
SUPPORT

Figure 5-43 Steel connector plates

Built-up header vs. single header (traditional vs. advanced framing). (See Table 5-1).

- Single headers used in combination with insulation are much more energy efficient, they create a cavity for insulation and use less framing material (see Figure 5-44).

- Many carpenters use built-up headers out of habit, not necessity, and end up wasting material and insulation value.

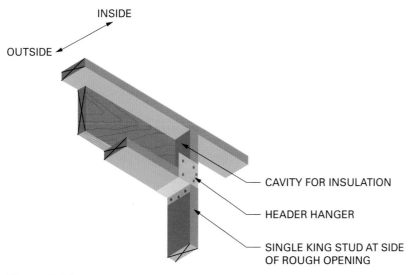

INSIDE

OUTSIDE

CAVITY FOR INSULATION

HEADER HANGER

SINGLE KING STUD AT SIDE
OF ROUGH OPENING

Figure 5-44 King stud with header hanger hardware

Three- or four-stud corner vs. two-stud corner with F-clips (traditional vs. advanced framing). (See Table 5-1).

- Two-stud corner saves one or two studs per outside corner but must be used with drywall clips to provide a method to fasten sheetrock. Furthermore, it may be difficult to apply some types of siding (see Figure 5-45).

- Two-stud corners are easier to insulate and will reduce thermal bridging.

- Three- of four-stud corners give more nailing surfaces for sheathing, corner boards (siding), and interior sheetrock.

- Some building codes may require a minimum of three studs at each exterior corner.

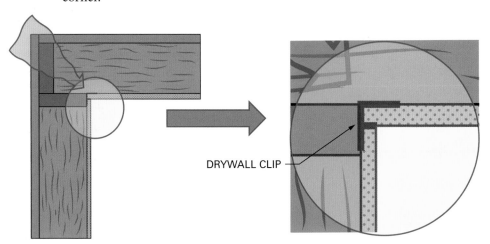

DRYWALL CLIP

Figure 5-45a Two-stud corner with drywall clip

DRYWALL CLIPS ARE PLACED AT REGULAR INTERVALS SUCH AS 16 OC VERTICALLY. THEY PROVIDE A SURFACE TO SCREW THE DRYWALL. IN THIS CASE, THE FIRST SHEET WILL BE FASTENED TO THE CLIPS ON THE RIGHT SIDE OF THE CORNER, THE SECOND SHEET WILL BE FASTENED TO THE STUDS ON THE LEFT WALL. DRYWALL CLIPS COME IN DIFFERENT SHAPES/TYPES, SOME "L" SHAPED, OTHERS "F" SHAPED.

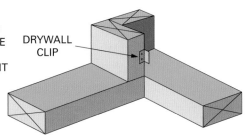

DRYWALL CLIP

Figure 5-45b Drywall support clips

Fewer design considerations vs. design and build with 2′ modules (traditional vs. advanced framing). (See table 5-1).

- Traditional framing is generally built using 2′ increments when considering width/length, but not when locating doors/windows.

- Advanced framing not only considers width/length but also attempts to locate windows/doors to use at least one of the 24″ OC framing members; some designers find this restrictive.

Fewer construction restrictions vs. may need to stack the framing (traditional vs. advanced framing). (See Table 5-1).

- When framing 24″ OC and single top plates are used, all structural members must be stacked over the top of each other (see Figure 5-46).

- To frame two-stories 24″ OC, 2 × 6 studs must be used.
- Stacking the framing takes time/effort to ensure proper alignment of framing members.

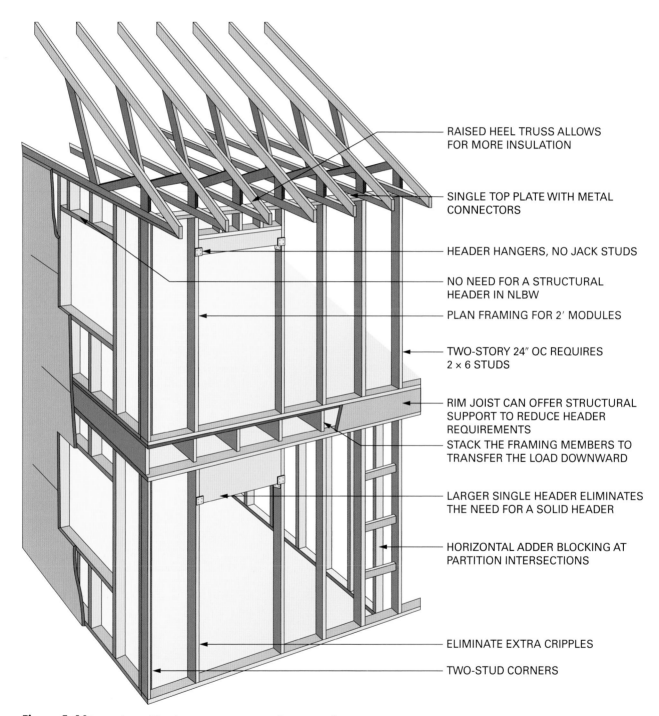

RAISED HEEL TRUSS ALLOWS FOR MORE INSULATION

SINGLE TOP PLATE WITH METAL CONNECTORS

HEADER HANGERS, NO JACK STUDS

NO NEED FOR A STRUCTURAL HEADER IN NLBW

PLAN FRAMING FOR 2′ MODULES

TWO-STORY 24″ OC REQUIRES 2 × 6 STUDS

RIM JOIST CAN OFFER STRUCTURAL SUPPORT TO REDUCE HEADER REQUIREMENTS

STACK THE FRAMING MEMBERS TO TRANSFER THE LOAD DOWNWARD

LARGER SINGLE HEADER ELIMINATES THE NEED FOR A SOLID HEADER

HORIZONTAL ADDER BLOCKING AT PARTITION INTERSECTIONS

ELIMINATE EXTRA CRIPPLES

TWO-STUD CORNERS

Figure 5-46 Some advanced framing concepts, a two-story house example

Notching and Boring

NLBW/P studs should not be notched greater than 40% of the stud width, and holes cannot be greater than 60% of stud width. Holes cannot be closer than 5/8″ from the edge of the stud.

LBW/P studs cannot be notched greater than 25% of stud width, and holes cannot be greater than 40% of the stud width. Holes cannot be closer than 5/8″ from the edge of the stud. As always, check your building code! See figures 5-47 & 5-48.

> *Holes can be bored up to 60% of the width of the stud if studs have been doubled. No more than two successive double studs can be so bored.*

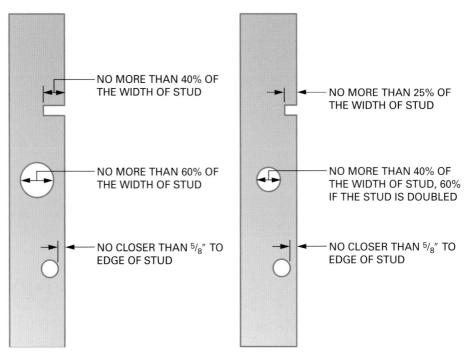

Figure 5-47 Rules for notching and boring interior non–load-bearing walls/partitions

Figure 5-48 Rules for notching/boring exterior walls and interior load-bearing walls

Framing Sequencing/Techniques

When assembling walls in typical platform construction, first the floor assembly is completed, including the sub-floor. Next, the walls are built. There are two preferred methods.

> **TIP**
>
> *Carpenters prefer, whenever possible, to assemble wall/partition sections including sheathing, before standing up the wall. This minimizes work from ladders and scaffolding.*

- **Method 1**—Some carpenters prefer to strike chalk lines on the sub-floor to identify all wall/partition locations, and then cut the wall plates. At this point, all wall layout is completed before assembling and erecting the wall sections.
- **Method 2**—Others prefer to lay out and build the walls one at a time, generally starting with the long load-bearing walls and finishing with the shorter partitions.

Both methods will achieve similar results and, as with other techniques, carpenters will good-naturedly argue the merits of methodology.

ASSEMBLY

Figure 5-49 illustrates the process of assembling a wall section. In order to speed construction, walls are generally built while lying on the deck, then stood up in manageable sections.

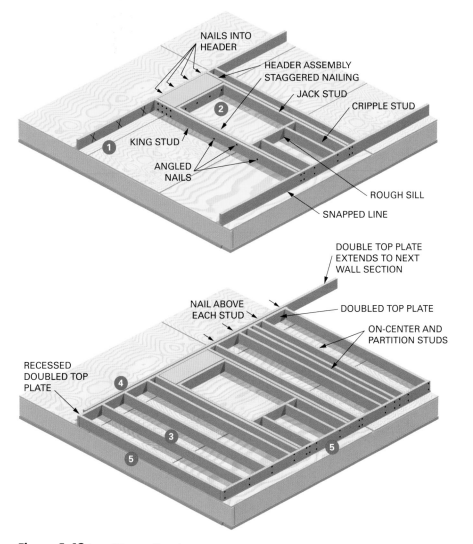

Figure 5-49 Assembling a wall section

CAUTION

Beware of windy conditions. A sheathed wall is heavy and has a large surface area that can catch the wind. This can cause injury or damage if dropped.

TIP

Some carpenters prefer to pre-assemble window sections and then simply drop them into the appropriate locations when building the walls.

TIP

Some carpenters prefer to crown all studs, placing the crown to the outside of the wall. Any studs with severe crowns should be set aside or cut for shorter components.

- **Step 1**—Lay out wall plates (see Figure 5-10 for more information). Do not cut bottom plates where there are door openings. They can be cut out after the walls are secured in place; this will help to maintain proper spacing and alignment.

- **Step 2**—Assemble any framing with openings such as windows (assembly order should allow end nailing as much as possible).

- **Step 3**—Install remaining studs (end nail through the plates).

When measuring from corner to corner, make sure to measure to the same relative places on each end. For example, measure from the bottom plate diagonally to the corner of the first top plate. Don't measure to the doubled plate because it may be offset on one end to allow for a tie-in with other walls.

- **Step 4**—Install doubled top plate, making sure to leave space to tie in the overlapping doubled plates.
- **Step 5**—Align framed wall to the chalk line on the deck and adjust it to be square (Figure 5-50).
- **Step 6**—Apply permanent bracing or sheathing to the wall (see Figure 5-52).

Squaring a Wall Section

Align the bottom plate of the wall to the chalk line on the deck, then measure from corner to corner across the diagonals (Figure 5-50). If the measurements are not the same, adjust the wall (by tapping one end of a top plate with a sledge hammer) until the diagonal measurements are identical. Finally, add sheathing or bracing to hold the wall square.

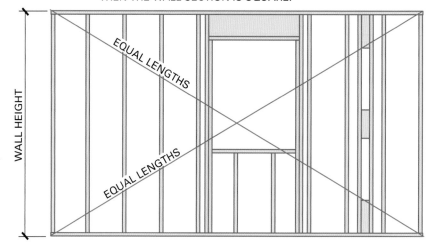

IF WALL STUDS ARE EQUAL LENGTH, WALL PLATES ARE EQUAL LENGTH, AND THE DIAGONALS ARE EQUAL, THEN THE WALL SECTION IS SQUARE.

Figure 5-50 Measure corner to corner to check for square

One can also calculate the theoretical lengths of the diagonals using the Pythagorean Theorem. These calculated values can also be used to measure the diagonals. See Chapter 2, "Squaring and Checking for Square" section.

The following is a more thorough explanation of the squaring procedure:

- **Step 1**—If not already done, strike a chalk line on the deck surface at the location where the wall will be placed.
- **Step 2**—Lightly tack the bottom plate of the wall to the floor along the inner edge (see Figure 5-51).
- **Step 3**—Square the wall as described earlier, then tack the outer edge of the top plate to the deck to hold the wall square.
- **Step 4**—Apply sheathing or bracing (see Figure 5-52).
- **Step 5**—Remove nails holding the top plate to the deck and stand up the wall. The nails in the bottom plate will pull out as the wall is raised into position.
- **Step 6**—Working from one end of the wall to the other, secure the wall along the chalk line by nailing through the bottom plate and sub-floor into the joists beneath. Plumb and brace the wall. See finished wall, Figure 5-53.

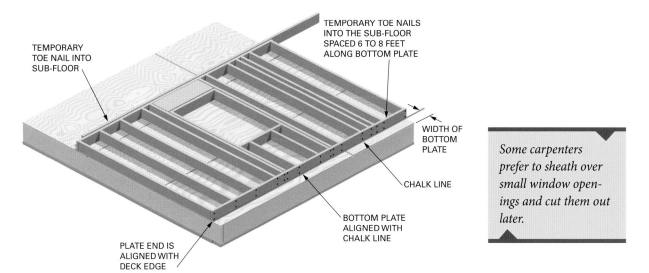

TEMPORARY TOE NAILS INTO THE SUB-FLOOR SPACED 6 TO 8 FEET ALONG BOTTOM PLATE

TEMPORARY TOE NAIL INTO SUB-FLOOR

WIDTH OF BOTTOM PLATE

CHALK LINE

BOTTOM PLATE ALIGNED WITH CHALK LINE

PLATE END IS ALIGNED WITH DECK EDGE

Some carpenters prefer to sheath over small window openings and cut them out later.

Figure 5-51 Temporarily attach wall to deck before applying sheathing

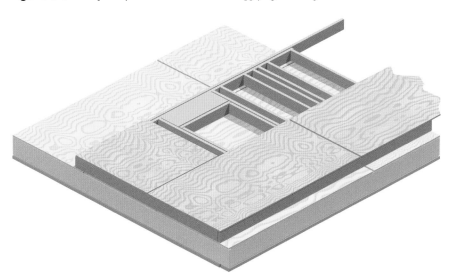

Figure 5-52 Applying sheathing

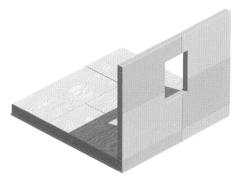

Figure 5-53 The sheathed wall

In some regions, if the sheathing is applied horizontally, it must have solid blocking behind the joints. If applied vertically, codes may require it to overlap the box header or include special hardware or strapping to unify the structure.

Sheathing

The wall can be sheathed before it is stood up or after, and there are different methods of applying sheathing (Figure 5-54). Note the nailing requirements table.

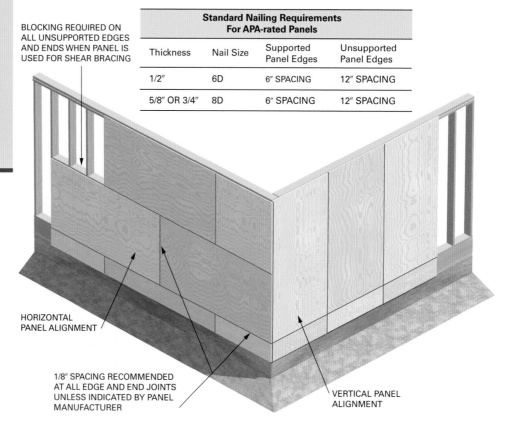

BLOCKING REQUIRED ON ALL UNSUPPORTED EDGES AND ENDS WHEN PANEL IS USED FOR SHEAR BRACING

Standard Nailing Requirements For APA-rated Panels			
Thickness	Nail Size	Supported Panel Edges	Unsupported Panel Edges
1/2"	6D	6" SPACING	12" SPACING
5/8" OR 3/4"	8D	6" SPACING	12" SPACING

HORIZONTAL PANEL ALIGNMENT

1/8" SPACING RECOMMENDED AT ALL EDGE AND END JOINTS UNLESS INDICATED BY PANEL MANUFACTURER

VERTICAL PANEL ALIGNMENT

Figure 5-54 Methods of installing APA-rated panel wall sheathing

Bracing

Permanent Bracing In colder climates, structural or non-structural insulated panels are often used/required as sheathing. If non-structural panels are used, permanent diagonal bracing must be integrated into the walls. Figure 5-55 illustrates a let-in brace. The let-in brace is typically a 1 × 4 and is installed flush with the surface of the studs. This is achieved by notching all of the studs and plates to accommodate the brace.

Another type of diagonal brace is a cut-in brace. This is made of several cut pieces, generally the same size framing member as the wall studs. The pieces are beveled on each end to fit between the studs and are aligned to transfer the forces from one piece to the next. A "kicker" is added at each end to further stabilize the brace and prevent studs from deflection (Figure 5-56).

Another type of brace (not shown) is a metal brace ("T"- or "L"-shaped in cross section). To use this type, strike a chalk line where the brace is needed and cut a notch (saw-blade width) to the appropriate depth across the studs and plates. Fasten the brace in the notch.

STUDS AND PLATES
ARE NOTCHED

CONTINUOUS
LET-IN BRACE

WALL HEIGHT

APPROXIMATELY EQUAL TO WALL HEIGHT

Figure 5-55 A let-in corner brace

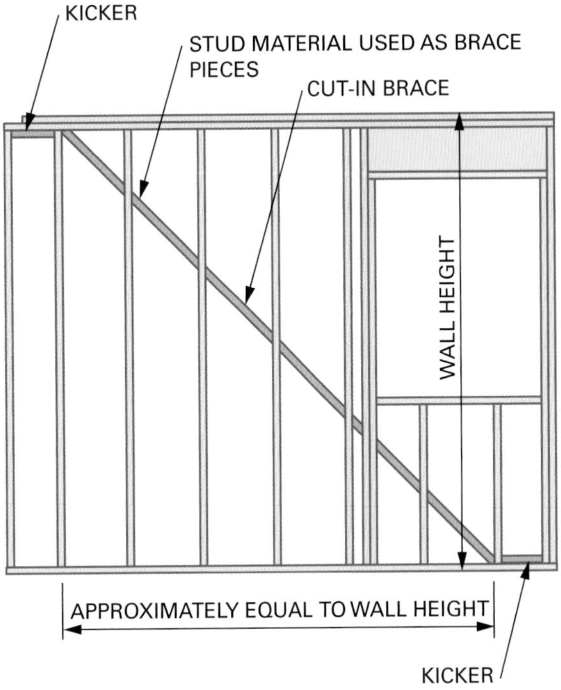

KICKER

STUD MATERIAL USED AS BRACE
PIECES

CUT-IN BRACE

WALL HEIGHT

APPROXIMATELY EQUAL TO WALL HEIGHT

KICKER

Figure 5-56 A cut-in brace

Temporary Bracing During the construction phase, temporary bracing is needed until other framing members are in place. Figure 5-57 shows a common method of temporarily bracing a wall section.

In earthquake-prone and high-wind areas, there may be more rigid requirements regarding permanent bracing and anchoring. Check local codes.

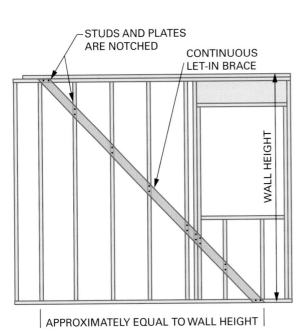

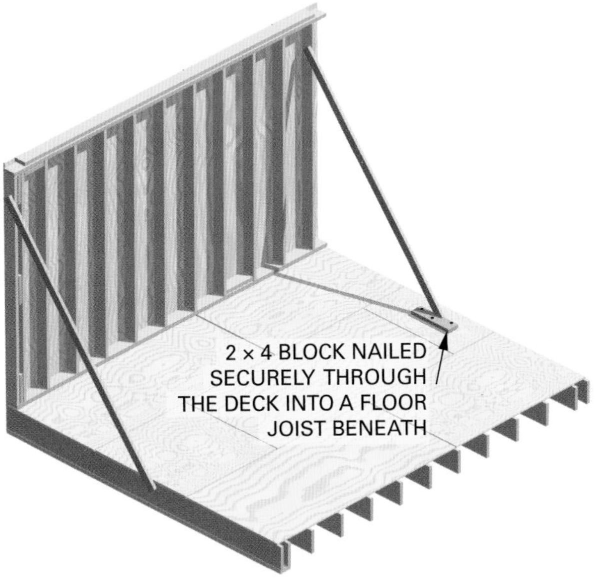

2 × 4 BLOCK NAILED
SECURELY THROUGH
THE DECK INTO A FLOOR
JOIST BENEATH

Figure 5-57 Temporary braces hold the frame erect during construction

Plumbing the Walls

> To "plumb" a wall is to make it stand perfectly vertical.

> Some studs are bowed badly enough that they need to be discarded or cut for use as smaller framing members.

The ultimate goal in plumbing a wall is to locate the top plate directly over the bottom plate. After standing up a wall, it must be fastened, plumbed, and braced. A quick way to plumb a wall is to use a level; always start at a corner. The longer the level, the more accurate it is. For wall framing applications a 4′ level is the minimum length to use; 6′ and 8′ levels are also available. Occasionally studs are bowed and placement of the level will affect whether or not the wall is truly vertical (plumb). In Figure 5-58-A, the level is held vertically; however, because the stud is bowed, the wall needs to lean farther to the right (out of plumb) to align with the level. Figure 5-58-B shows the reverse—because of the position of the level, the same wall needs to lean to the left for the level to read plumb. Using the level as shown in both "A" and "B" in Figure 5-58 will produce an out-of-plumb wall. If the level were positioned in the center as in "C" in Figure 5-58, the carpenter just might get lucky and set the wall plumb; however, the level will tend to rock back and forth in this position and the carpenter will not know the exact location to take the reading. If an 8′ level is not available, one accurate method is to use a straight stud with a block attached to each end. This will allow the carpenter to plumb from the top plate to the bottom plate regardless of how bowed the stud is. Once the wall is plumb, it can be braced. Keep in mind that unsheathed, unbraced walls need to be plumbed and braced in both directions.

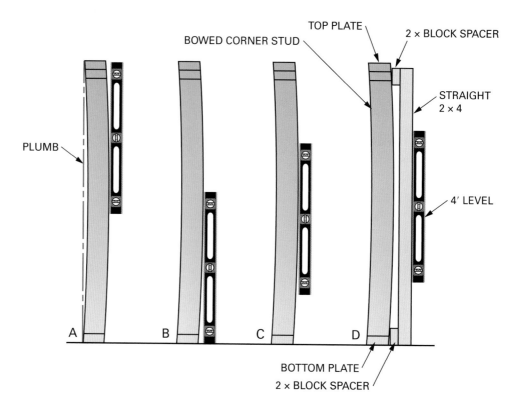

NOTE: IN REALITY MANY USABLE STUDS WILL HAVE A SLIGHT BOW AND THIS TECHNIQUE WILL INSURE A PLUMB WALL. IDEALLY BOWED STUDS SHOULD NOT BE USED.

Figure 5-58 Plumbing a wall/partition

Straightening a Wall

Often a long wall needs intermediate bracing for support and to help straighten it. To check if a long wall is straight, refer to Figure 5-59. After placing blocks at either end of the top of a wall, string a line (tight) along the top of the wall (a mason line works well) and measure in several places between the string and the top plate; if the measurements are different than the thickness of the end blocks, then adjust the wall accordingly. This may be done by using bracing at various points along the wall.

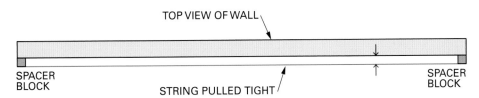

Figure 5-59 Stringing a wall

If the wall bows inward (to the right), place a diagonal brace as shown on the right side of Figure 5-57. If the wall is leaning outward, a similar brace (on the left) from the wall to a stake in the ground will also work. More than one brace will be necessary on a long wall. Figure 5-60 illustrates a simple two-piece brace that can be used to lever the wall in either direction.

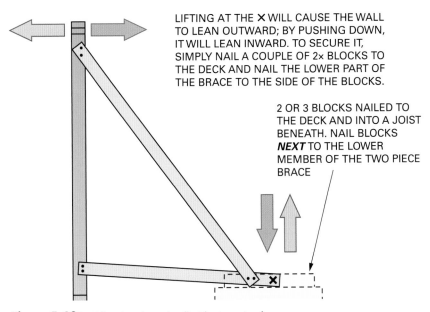

Figure 5-60 Straightening a bowed wall with a two-piece brace

Working with Odd Angles

Often carpenters find themselves having to build a structure requiring angles other than 90°. One method of determining the angle needed is to know the number of sides a structure has. A four-sided structure that is built with straight cuts (90° cuts) is represented by "A" in Figure 5-61. There are 360° in a polygon. To find the degree angle on each corner, simply divide the number of sides into 360°. In "A" of Figure 5-61, 360°/4 = 90°. However, what if the sides are put together with miter joints? There are still four sides but there are *two* angles per side. To solve this, simply divide 360° by the number of angles. For example, 360°/8 = 45°, four pieces with an angle on each end. This is represented by "B" in Figure 5-61.

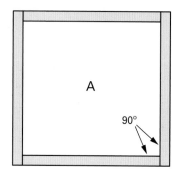
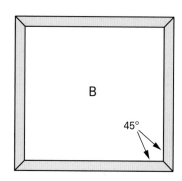
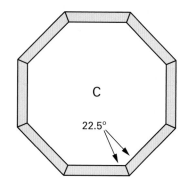

Figure 5-61 Working with odd angles

What if an octagon (eight-sided) structure were being built? 360°/16 = 22.5°. See "C" in Figure 5-61.

Remember, 8 corners and 2 angles per side = 16 angles.

Table 5-2 shows the angles of some common shapes.

Table 5-2 Table of angles

Shape	Number of Sides	Angle of Each Piece When Mitered
Square	4	45°
Pentagon	5	36°
Hexagon	6	30°
Octagon	8	22.5°
Decagon	10	18°
Dodecagon	12	15°

CHAPTER 6

RAFTERS/ ROOF FRAMING

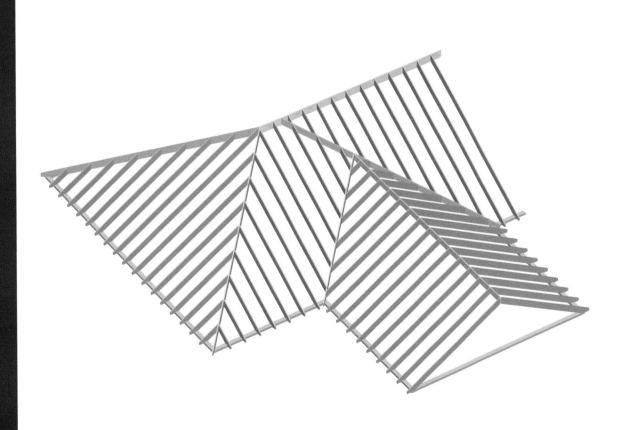

RAFTERS/ROOF FRAMING

This chapter covers several complex concepts. In order to follow the terminology and processes, it is advisable to read the chapter in the exact order it is presented. There are some concepts covered early in the chapter that are not covered again later.

Different Roof Types

There are several distinct roof types (see Figure 6-1). Some were developed for specific practical reasons and others for stylistic reasons. The most common is the gable roof. One style not shown is the flat roof. It generally has a very shallow slope but is not quite flat.

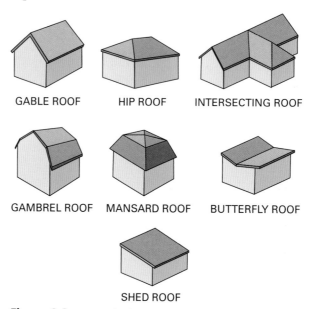

GABLE ROOF   HIP ROOF   INTERSECTING ROOF

GAMBREL ROOF   MANSARD ROOF   BUTTERFLY ROOF

SHED ROOF

Figure 6-1 Many roof styles can be used for residential buildings

In Chapter 6 there are three glossaries. Terms are defined in groups as they are presented.

Basic Roof Components Glossary

For the following terms, see Figure 6-2.

Bird's mouth—A notch cut in the rafter so it can rest on the top plate of the wall.

Collar tie—Connects opposing pairs of rafters, generally 1/3 of the span of the building.

Fly rafter—Also known as a barge rafter or rake rafter. The rafter forming the overhang on the gable end of a building.

Lookouts—Framing members that help support the fly rafters.

Rafter—Framing members that make up the roof.

Rafter square—A square with embossed tables that help with rafter calculations (see Figure 6-72).

Rafter tail—The portion of the rafter that overhangs the structure.

Ridge—The uppermost framing member. Rafters generally terminate and attach to the ridgeboard.

Ridge post—Framing member that supports the ridge when a structural ridge-board is used.

Run—Measured from the outside wall to the center of the ridge.

Soffit—Found underneath the rafter tail between the fascia and the wall of the structure.

Sub-fascia—Connects all of the rafter tails together and acts as a base for the finished fascia.

THE GABLE ROOF

Rafters used to frame a gable roof are called *common rafters*; on a basic gable roof all rafters are identical.

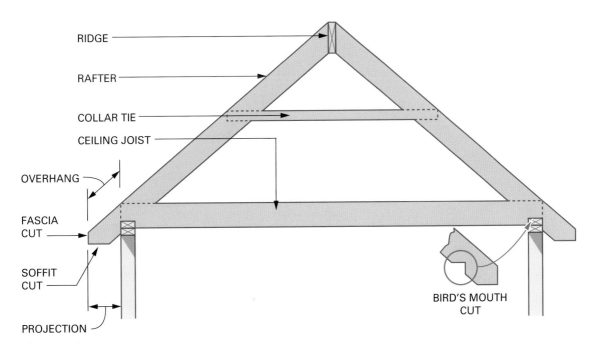

Figure 6-2 Gable roof components

Figure 6-3 shows the relationship between components as they sit on a framed wall.

Lookouts may be needed to help support the fly rafter (see Figure 6-4). The top of the lookouts are flush with the top of the rafters and help support the roof sheathing.

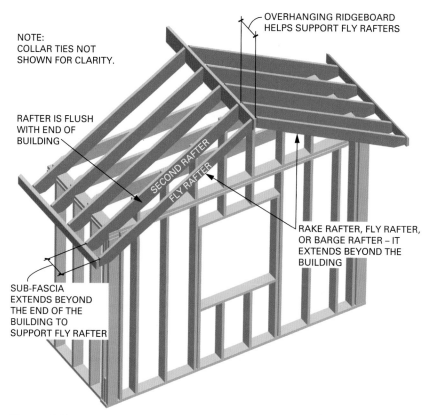

NOTE:
COLLAR TIES NOT
SHOWN FOR CLARITY.

OVERHANGING RIDGEBOARD
HELPS SUPPORT FLY RAFTERS

RAFTER IS FLUSH
WITH END OF
BUILDING

SECOND RAFTER

FLY RAFTER

RAKE RAFTER, FLY RAFTER,
OR BARGE RAFTER – IT
EXTENDS BEYOND THE
BUILDING

SUB-FASCIA
EXTENDS BEYOND
THE END OF THE
BUILDING TO
SUPPORT FLY RAFTER

Figure 6-3 Gable roof with fly rafters

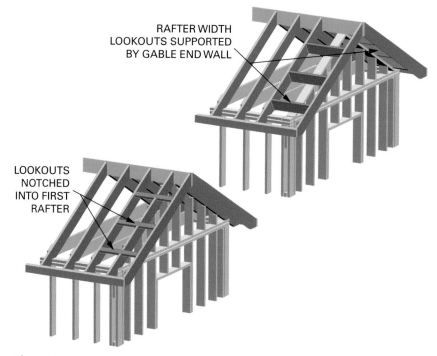

RAFTER WIDTH
LOOKOUTS SUPPORTED
BY GABLE END WALL

LOOKOUTS
NOTCHED
INTO FIRST
RAFTER

Fly rafters are the same length as common rafters but generally not as wide. Common rafters may be 2 × 10 or larger and the fly rafters only 2 × 6 or 2 × 4; therefore, they need extra support (lookouts) to keep them from sagging.

Figure 6-4a Two styles of lookouts can support the rake overhang

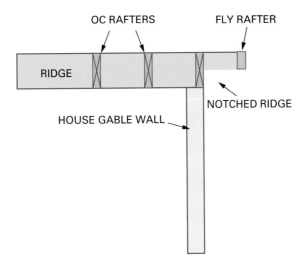

Figure 6-4b A common ridge/fly rafter detail

Typically, the fly rafter is smaller than the other rafters. If large rafters are used, the fly rafter is often made from a 2 × 6 or even a 2 × 4.

There are several methods of framing the ridge/fly rafter intersection. One method is to lean the fly rafters against each other at their tips. When using this method, contact with the ridge will be unnecessary. Another method is to cantilever the ridge outward and notch it so that the overhanging portion of the ridge will be flush with the bottom of the fly rafter. This way, the ridge will help to carry the overhanging rake edge, but not obstruct the soffit.

Rafter Calculation Terminology Glossary

These terms will be described in detail before attempting calculations of rafter lengths. See Figures 6-5 and 6-6.

Line length—Measured along the edge of the rafter from the center of the ridge to the back (plumb line) of the bird's mouth cut (measured along the dashed line shown in Figure 6-6).

Overhang—The sloped distance measured along the rafter tail that extends beyond the side of the structure.

Pitch—A statement (fraction) of the steepness of a roof, determined by comparing the unit rise to the unit span. For example, if the unit rise is 8″ and the unit span 24″, the pitch will be stated as an "8/24 pitch." However, the fraction must be simplified, so this is referred to as a "⅓ pitch."

Projection—The horizontal distance measured from the side of the structure to the front of the fascia.

Slope—A statement of steepness of a roof determined by comparing unit rise to unit run. For example, if the unit rise for a common rafter is 8″, then for every unit of run (12″ horizontal) the roof will rise 8″. Carpenters call this an "8 on 12 slope." Or simply an 8/12 slope.

Total rise—The amount a rafter rises from the top plate to the center of the ridge.

Total run—Half of the span, which is half of the width of the building. Or more specifically, the horizontal distance from the center of the ridge to the outside of the exterior wall.

Total span—The width of the building.

Unit length—This is the hypotenuse of the unit triangle (Figure 6-5).

Unit rise—The amount a rafter rises per unit of run. Unit rise when compared to the unit run determines the steepness of the slope of the roof.

Unit run—The unit of run for a *common rafter* is **always 12″**. It is thought of terms of 1 unit (12″ = 1 unit). The unit of run for a *hip rafter* is **16.97″**.

Unit span—Always 24″ for common rafters—twice the unit of run.

Unit triangle—A small right triangle, generally shown on an elevation view of the roof. The unit triangle comprises the unit run, unit rise, and unit length. The unit triangle can be used to help with rafter calculations (Figure 6-5).

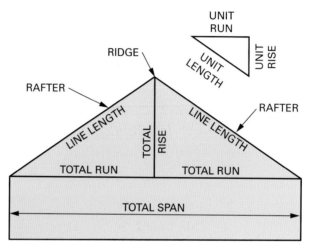

Figure 6-5 The terms associated with roof theory

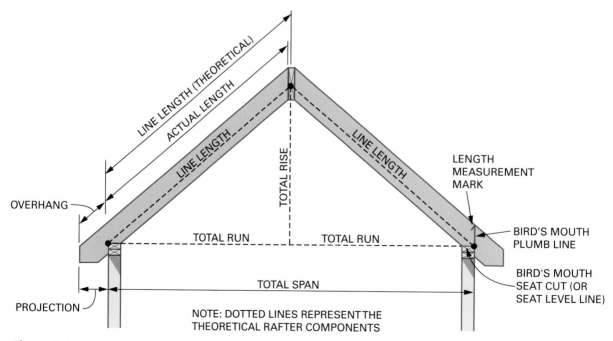

Figure 6-6 The line length is theoretical, measured before the rafter is shortened to allow for the ridge

Calculating Unit Lengths

The small triangle in the upper right of Figure 6-5 is called the unit triangle, also shown in Figure 6-7. It is found on elevation drawings of the roof and is proportional to the shape of the roof. On a plan, in place of the term unit rise, there will be a number, such as 8. This number means that for every 12″ of run there are 8″ of rise. A carpenter may call this an "8 on 12 roof slope." Or an 8/12 slope.

Figure 6-7 The unit triangle may be represented either way

The unit length is the diagonal distance (hypotenuse) between the two legs of the unit triangle (see Figure 6-5). If one leg of the triangle is 12″ and the other is 8″, then using the Pythagorean Theorem:

$$a^2 + b^2 = c^2 \text{ (} c \text{ being the hypotenuse [unit length],}$$
$$\text{and } a \text{ and } b \text{ being the legs).}$$

$$8^2 + 12^2 = c^2$$

$$64 + 144 = c^2$$

$$208 = c^2$$

$$\sqrt{208} = 14.42 = c$$

Many framing squares have tables showing some of the unit lengths and other information (see Figure 6-72). Table 6-1 shows some of the frequently used unit rises and their calculated unit lengths. For this part of the chapter, only the *Common Rafter* Unit Length values will be used, as opposed to Hip or Valley rafters.

Table 6-1 Table of unit lengths

Unit Rise	Common Rafter Unit Length	Hip/Valley Rafter Unit Length
3	12.37	17.23
4	12.65	17.44
5	13	17.69
6	13.42	18
7	13.89	18.36
8	14.42	18.76
9	15	19.21
10	15.62	19.70
11	16.28	20.22
12	16.97	20.78

Carpenters often mistakenly exchange the term pitch for the term slope. Slope is a relationship that compares the unit rise to the unit run (for example: 6/12 slope). Pitch, on the other hand, is the relationship of the unit rise to the unit span, which is 24″. For example, a 6/24 pitch is a fraction that must be simplified to its simplest form, which is 1/4. Therefore, a 6/12 slope = 1/4 pitch.

In the case where a roof with an intermediate unit rise is needed, such as 4½ on 12, use the Pythagorean Theorem to calculate the unit length.

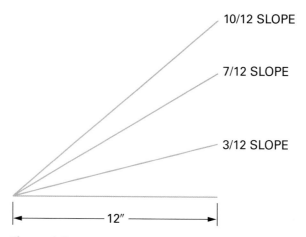

Figure 6-8 Different unit rises showing steepness of roof slope

Figure 6-8 shows the relative steepness of different roof slopes. It is possible to frame very steep roofs such as a church steeple with slopes of 40/12 or even steeper.

Calculating Common Rafter Lengths

Figure 6-9 shows another view of a rafter as it is oriented on a building.

Run is defined as half of the span (or half of the width of the building). Using the formula **unit length × run = line length**, a rafter (line) length can be calculated.

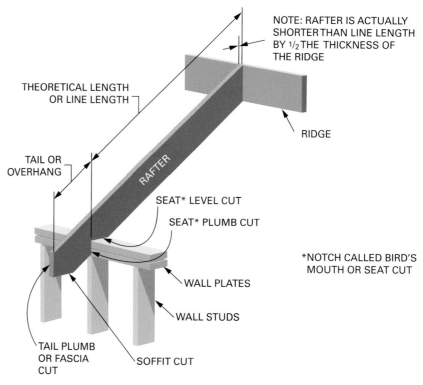

> *The line length does not include the rafter tail. It only includes the rafter as measured from the center of the ridge to the seat plumb cut (back of the bird's mouth).*

Figure 6-9 Names of the cuts and lines of a common rafter

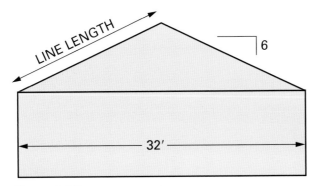

Figure 6-10 Find the line length

Important Note: When using this method of calculation, the unit length should be indicated in decimal inches and the run indicated in decimal feet (thought as units of run).

Example 1 (Figure 6-10): A building has a span of 32′ and the roof has a unit rise of 6 (6/12 slope). What is the line length of the rafter?

Unit rise = 6; therefore, unit length = 13.42 (Table 6.1). Span = 32′; therefore, run = 16′ (run = ½ span).

$$\text{Unit length} \times \text{run} = \text{line length}$$
$$13.42 \times 16 = 214.72'' = 214\tfrac{3}{4}''$$

Example 2: If a building has a span of 28′ 5″ (run = 14′ 2½″), and a unit rise of 11 (unit length = 16.28), what is the line length?

$$16.28 \times 14.208' = 231.306'' = 213\tfrac{5}{16}''$$

Calculating Line Length and Overhang (Total Rafter Length)

Rafters often overhang the edge of the building, as shown in Figure 6-11. Using the given projection dimension (16″) and the unit rise (10″), the overhang can be calculated.

Important Note: Think of the overhang as a small rafter and use the projection dimension as the run. Therefore:

$$\textbf{Unit length} \times \textbf{projection} \text{ (in feet)} = \textbf{overhang}$$

Example 1: Unit rise = 10 (unit length = 15.62). Span = 25′ (run = 12.5′). First, calculate the line length:

$$\text{Unit length} \times \text{run} = \text{line length}$$

therefore:

$$15.62 \times 12.5' = \textbf{195.25''} = 195\tfrac{1}{4}''$$

Next calculate the overhang using the formula above. First change the projection from inches to feet by dividing the projection dimension by 12:

$$16'' \div 12'' = 1.333' = \text{projection in feet}$$
$$\text{Unit length} \times \text{projection} = \text{overhang}$$
$$15.62 \times 1.333 = \textbf{20.82''} = 20\tfrac{13}{16}''$$

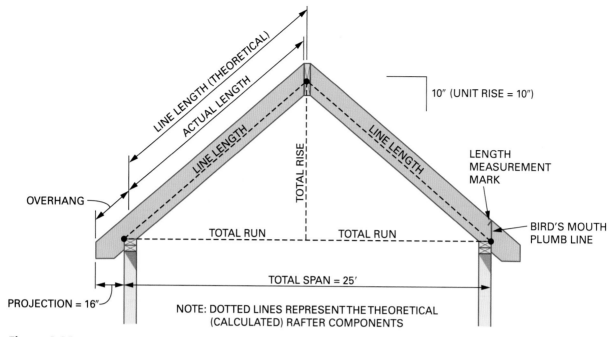

Figure 6-11 Calculate the total rafter length

This following step will help the carpenter determine what length lumber is needed:

Line length + overhang = total rafter length

$$195.25'' + 20.82'' = 217.07'' = \textbf{217}\frac{1}{16}''$$

Example 2: If the unit rise is 4 (unit length = 12.65) and the projection is 22″ (1.833′), how long is the overhang?

$$12.65 \times 1.833' = 23.187'' = \textbf{23}\frac{3}{16}''$$

Calculating Total Rise

Using the unit triangle, the total *rise* of the rafter can be calculated (see Figure 6-11).
Use the following formula: **Unit rise × run = total rise.**

The unit rise is represented in inches, just as it is shown on the unit triangle. The run should be in feet. From Figure 6-11, the span = 25′; therefore the run = 12.5, and the unit rise = 10″.

$$10'' \times 12.5 = \textbf{125}'' = \textbf{total rise}$$

COMMON RAFTER LAYOUT

The rafter shown in Figure 6-12 displays all markings and dimension lines needed to lay out and cut the rafter. Notice that all vertical lines (plumb lines) are parallel to each other (A, B, D, and G), and all horizontal lines are square to

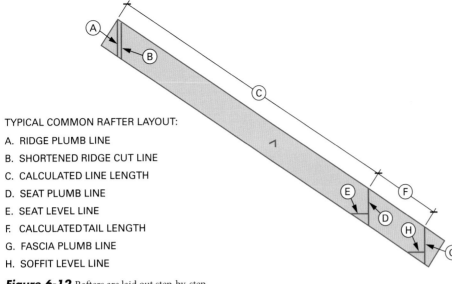

TYPICAL COMMON RAFTER LAYOUT:

A. RIDGE PLUMB LINE

B. SHORTENED RIDGE CUT LINE

C. CALCULATED LINE LENGTH

D. SEAT PLUMB LINE

E. SEAT LEVEL LINE

F. CALCULATED TAIL LENGTH

G. FASCIA PLUMB LINE

H. SOFFIT LEVEL LINE

Figure 6-12 Rafters are laid out step-by-step

> *The total rise, when added to the wall height, will help to determine the overall height of the building. Some building codes have maximum height restrictions. Furthermore, knowing this height can help the carpenter to identify the type of lifts, scaffolding, and ladders that will be needed. It is good practice to be able to determine this height before starting construction.*

the plumb lines (E and H). The inverted "V" is a crown mark that is placed on the board before layout starts; the tip of the mark indicates the direction of the crown. As in floor framing, the crown should be placed upward as shown in Figure 4-13.

The following are two methods of laying out a common rafter:

- **Method 1**—The *calculation method* uses the calculated line length and calculated overhang.

- **Method 2**—The *step-off method* requires use of a square to draw repeated markings of the number of units of run and the units of rise for the building.

Calculation Method of Rafter Layout (Method 1)

For a pictorial example of rafter layout, skip ahead to Figure 6-20.

- **Step 1**—Crown the board and face the crown away.

- **Step 2**—Place the framing square on the board as shown in Figure 6-13. The unit rise number, on the tongue of the square, should just touch the edge of the board. The unit run number (12″) on the body of the square should also align with the edge of the board. Draw a line along the tongue of the square. This is called the *ridge plumb line* (see Figure 6-12, line A).

- **Step 3**—Measure at right angles to the ridge plumb line ½ of the thickness of the ridge and draw another plumb line. This is done to allow space for the ridgeboard (see Figure 6-14). This shortened (second) line will be the line that is eventually cut (see Figure 6-12, line B).

- **Step 4**—Calculate and then measure the line length (Figure 6-15). Measure from the first plumb line drawn in Step 2 and mark this point on the top *edge* of the board (see also Figure 6-12, dimension C).

> *It may be necessary to place the square along the top edge of the board or along the bottom edge of the board. Either method will give the same result but one method may make it easier to draw the entire plumb line. Diagrams in this book will show both methods of square placement.*

This is not a book on tool usage but the speed square does bear mentioning. The speed square is a versatile triangular-shaped tool and can be carried in a tool belt. It can act as a square, a saw guide, and it has unit rise notations, allowing the carpenter to lay out rafters. There are also degree notations allowing angle measurement and layout.

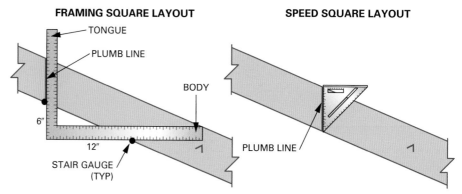

NOTE:
WHEN USING STAIR GAUGES, ALIGN THE UNIT RISE AND UNIT RUN ALONG THE EDGE OF THE RAFTER, THEN PLACE THE GAUGES ON THE SQUARE. DUE TO THEIR SHAPE, THE GAUGES MAY NOT ALIGN EXACTLY ON THE UNIT RISE AND UNIT RUN NUMBERS.

Figure 6-13 Marking the ridge plumb line

- **Step 5**—Draw another plumb line from the point marked in Step 4 (Figure 6-15). This represents the seat plumb line (see also Figure 6-12, line D).
- **Step 6**—Using a square, mark the seat level line (Figure 6-16). Make sure to leave at least 2/3 of the board intact above the intended seat level line; some

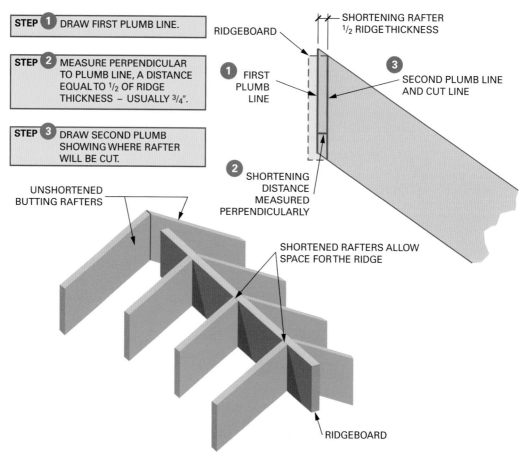

Figure 6-14 Shortening a rafter that abuts a ridgeboard

FRAMING SQUARE LAYOUT

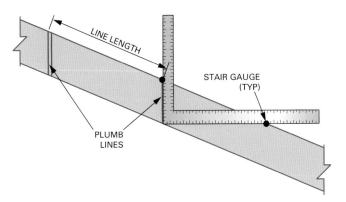

LINE LENGTH

STAIR GAUGE (TYP)

PLUMB LINES

> The seat level line and the lower portion of the seat plumb line define the "bird's mouth." The terms "bird's mouth cut" and "seat cut" are often used interchangeably.

NOTE:
SQUARE CAN BE ALIGNED WITH TOP EDGE OF BOARD AS SHOWN HERE, OR WITH THE BOTTOM EDGE OF THE BOARD AS SHOWN IN FIGURE 6-13. WHICHEVER WAY IS CHOSEN IS GENERALLY UTILIZED THROUGHOUT THE LAYOUT.

SPEED SQUARE LAYOUT

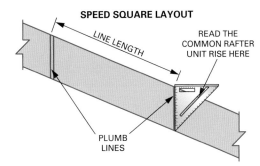

LINE LENGTH

READ THE COMMON RAFTER UNIT RISE HERE

PLUMB LINES

Figure 6-15 Marking the seat plumb line

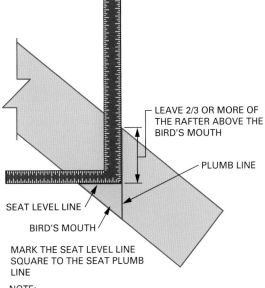

LEAVE 2/3 OR MORE OF THE RAFTER ABOVE THE BIRD'S MOUTH

PLUMB LINE

SEAT LEVEL LINE

BIRD'S MOUTH

MARK THE SEAT LEVEL LINE SQUARE TO THE SEAT PLUMB LINE

NOTE:
IF USING STAIR GAUGES, MARK LEVEL LINES (SEAT & SOFFIT) ALONG THE BODY OF THE SQUARE AND THE PLUMB LINES ALONG THE TONGUE

Figure 6-16 Marking the seat level line

codes allow the seat to be a minimum of 1½″, the seat can be as long as the wall plate's width but should not be longer (see also Figure 6-12, line E).

- **Step 7**—Calculate the overhang to determine the rafter tail length (Figure 6-17). To mark this point, measure from the seat plumb line (see also Figure 6-12, dimension F).

- **Step 8**—Draw a plumb line through the point marked in Step 7. This line is called the fascia plumb line. The board may not be long enough to allow the square to be held in the same manner as before, so it may have to be marked from the upside-down position as shown in Figure 6-18 (also see Figure 6-12, line G).

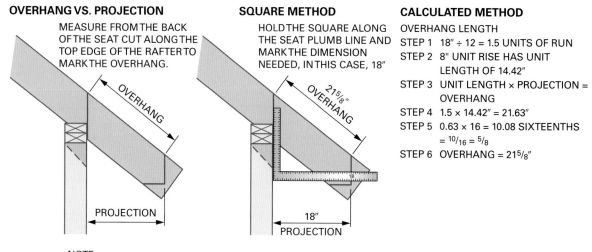

OVERHANG VS. PROJECTION

MEASURE FROM THE BACK OF THE SEAT CUT ALONG THE TOP EDGE OF THE RAFTER TO MARK THE OVERHANG.

SQUARE METHOD

HOLD THE SQUARE ALONG THE SEAT PLUMB LINE AND MARK THE DIMENSION NEEDED, IN THIS CASE, 18″

CALCULATED METHOD

OVERHANG LENGTH

STEP 1 18″ ÷ 12 = 1.5 UNITS OF RUN

STEP 2 8″ UNIT RISE HAS UNIT LENGTH OF 14.42″

STEP 3 UNIT LENGTH × PROJECTION = OVERHANG

STEP 4 1.5 × 14.42″ = 21.63″

STEP 5 0.63 × 16 = 10.08 SIXTEENTHS = $^{10}/_{16}$ = $^{5}/_{8}$

STEP 6 OVERHANG = 21$^5/_8$″

NOTE:
THE RAFTER TAILS ARE GENERALLY LAID OUT BEFORE CONSTRUCTION; THE WALLS ARE SHOWN ONLY TO DEMONSTRATE THE RELATIVE LOCATION AND ORIENTATION OF THE COMPONENTS.

Figure 6-17 Laying out the tail using the projection and overhang methods

Many carpenters use a hybrid layout technique and instead of calculating the overhang, they lay it out using the projection measurement (see right side of Figure 6-17). This technique is borrowed from the step-off method of rafter layout (discussed below). It is popular due to its speed and simplicity.

FRAMING SQUARE LAYOUT

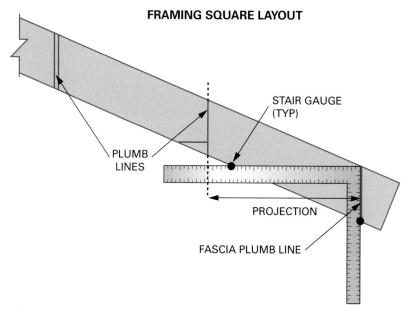

STAIR GAUGE (TYP)

PLUMB LINES

PROJECTION

FASCIA PLUMB LINE

Figure 6-18 Marking the fascia plumb line

- **Step 9**—Measure the desired length of the fascia plumb line (from the top edge of the rafter) and square back from this line to mark the soffit cut line (see Figure 6-19a). This line location depends on the size of material being used for the fascia and the soffit construction. The soffit line is also represented in Figure 6-12, line H.

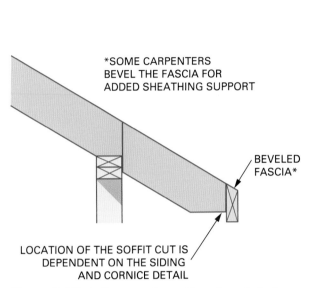

*SOME CARPENTERS BEVEL THE FASCIA FOR ADDED SHEATHING SUPPORT

BEVELED FASCIA*

LOCATION OF THE SOFFIT CUT IS DEPENDENT ON THE SIDING AND CORNICE DETAIL

Figure 6-19a Soffit cuts are often at right angles to the fascia

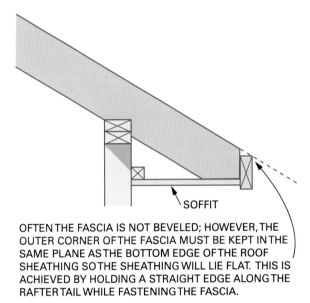

SOFFIT

OFTEN THE FASCIA IS NOT BEVELED; HOWEVER, THE OUTER CORNER OF THE FASCIA MUST BE KEPT IN THE SAME PLANE AS THE BOTTOM EDGE OF THE ROOF SHEATHING SO THE SHEATHING WILL LIE FLAT. THIS IS ACHIEVED BY HOLDING A STRAIGHT EDGE ALONG THE RAFTER TAIL WHILE FASTENING THE FASCIA.

Figure 6-19b Placement of a non-beveled fascia

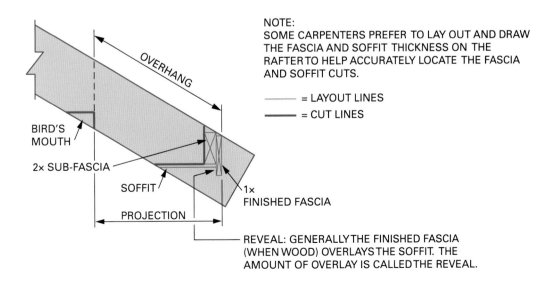

OVERHANG

NOTE:
SOME CARPENTERS PREFER TO LAY OUT AND DRAW THE FASCIA AND SOFFIT THICKNESS ON THE RAFTER TO HELP ACCURATELY LOCATE THE FASCIA AND SOFFIT CUTS.

——— = LAYOUT LINES
▬▬▬ = CUT LINES

BIRD'S MOUTH

2× SUB-FASCIA

SOFFIT

1× FINISHED FASCIA

PROJECTION

REVEAL: GENERALLY THE FINISHED FASCIA (WHEN WOOD) OVERLAYS THE SOFFIT. THE AMOUNT OF OVERLAY IS CALLED THE REVEAL.

Figure 6-19c Marking the fascia and soffit details

Figure 6-20 describes the procedure again without the detailed explanations.

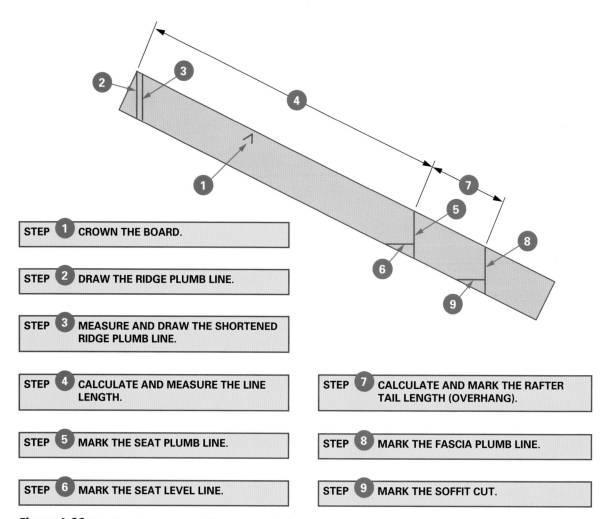

STEP **1** CROWN THE BOARD.

STEP **2** DRAW THE RIDGE PLUMB LINE.

STEP **3** MEASURE AND DRAW THE SHORTENED RIDGE PLUMB LINE.

STEP **4** CALCULATE AND MEASURE THE LINE LENGTH.

STEP **5** MARK THE SEAT PLUMB LINE.

STEP **6** MARK THE SEAT LEVEL LINE.

STEP **7** CALCULATE AND MARK THE RAFTER TAIL LENGTH (OVERHANG).

STEP **8** MARK THE FASCIA PLUMB LINE.

STEP **9** MARK THE SOFFIT CUT.

Figure 6-20 Using the calculation method for common rafter layout

Step-off Method of Common Rafter Layout (Method 2)

This technique eliminates the need for calculations; however, due to repetitive markings, maintaining accuracy while marking is of high importance. Stair gauges can reduce the error and make this process much easier (see Figure 6-13).

Follow the procedure in Figures 6–21 and 6–22 to perform an $8/12$ rafter layout using the step-off method.

Figure 6-23 illustrates why the step-off method works.

STEP 1 ALIGN 8″ AND 12″ OF THE SQUARE ON THE TOP EDGE OF THE RAFTER AND ATTACH THE STAIR GUAGES.

STEP 2 SLIDE SQUARE LEFT TO THE END OF THE BOARD AND CHECK 8″ AND 12″ ALIGNMENT.

STEP 3 HOLD AND MARK THE PLUMB LINE ALONG THE TONGUE.

STEP 4 STILL HOLDING THE SQUARE, MAKE A VERTICAL TIC MARK UNDER 12″ OF THE BLADE.

STEP 5 SLIDE SQUARE TO THE RIGHT UNTIL THE PREVIOUS TIC MARK ALIGNS WITH THE OUTSIDE-EDGE TONGUE. CHECK THE 8″ AND 12″ ALIGNMENTS.

STEP 6 HOLD SQUARE AND MARK ANOTHER TIC MARK UNDER 12″ OF THE BLADE.

STEP 7 REPEAT UNTIL ALL STEP OFF MARKS HAVE BEEN DRAWN. DRAW AS MANY AS THERE ARE UNITS OF RUN IN THE BUILDING.

STEP 7A SEE FIGURE 6-22 IF THERE ARE NOT AN EVEN NUMBER OF UNITS OF RUN.

STEP 8 LAST STEP IS MARKED WITH A FULL PLUMB LINE TO INDICATE THE SEAT PLUMB LINE.

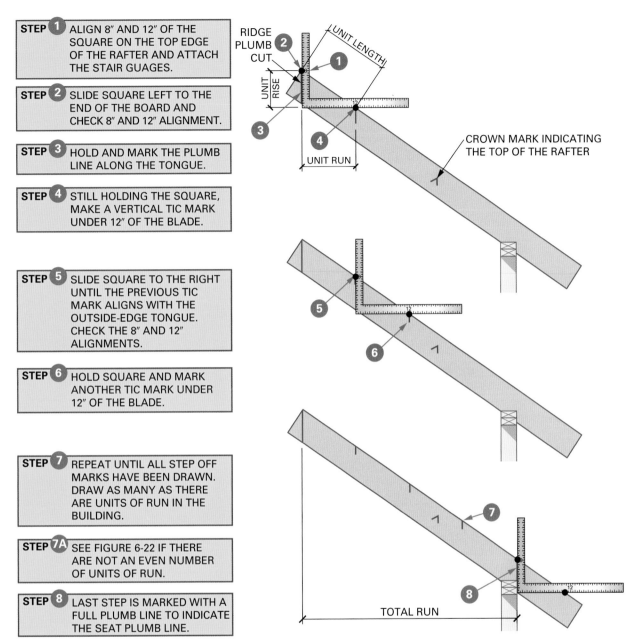

RIDGE PLUMB CUT

UNIT LENGTH

UNIT RISE

UNIT RUN

CROWN MARK INDICATING THE TOP OF THE RAFTER

TOTAL RUN

NOTE:
BEFORE CUTTING, DON'T FORGET TO MARK THE SECOND (SHORTENED) PLUMB LINE TO ACCOMMODATE THE RIDGE.

Figure 6-21 Using the step-off method for layout of an 8 on 12 rafter

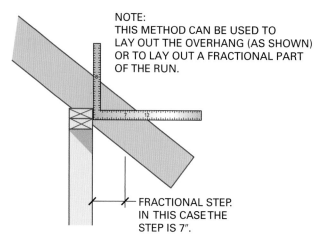

NOTE:
THIS METHOD CAN BE USED TO
LAY OUT THE OVERHANG (AS SHOWN)
OR TO LAY OUT A FRACTIONAL PART
OF THE RUN.

FRACTIONAL STEP.
IN THIS CASE THE
STEP IS 7".

Figure 6-22 Laying out a fractional part of a run for a common rafter

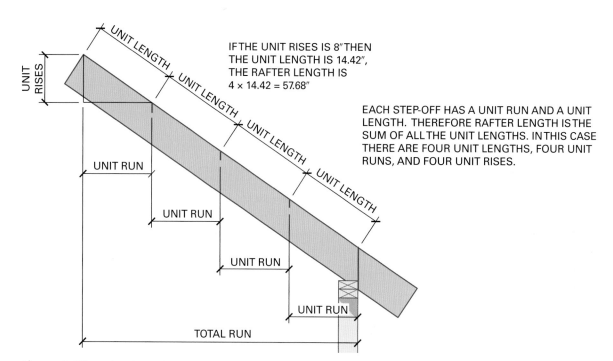

IF THE UNIT RISES IS 8" THEN
THE UNIT LENGTH IS 14.42",
THE RAFTER LENGTH IS
$4 \times 14.42 = 57.68"$

EACH STEP-OFF HAS A UNIT RUN AND A UNIT
LENGTH. THEREFORE RAFTER LENGTH IS THE
SUM OF ALL THE UNIT LENGTHS. IN THIS CASE
THERE ARE FOUR UNIT LENGTHS, FOUR UNIT
RUNS, AND FOUR UNIT RISES.

Figure 6-23 Unit length is the hypotenuse of the right triangle included in each step-off

Types of Rafter Tails

Different tail types depend on the house style. Figure 6-24 shows some different ways to shape the rafter tails.

Some structures do not have any overhang, and still others display decoratively cut exposed rafter tails.

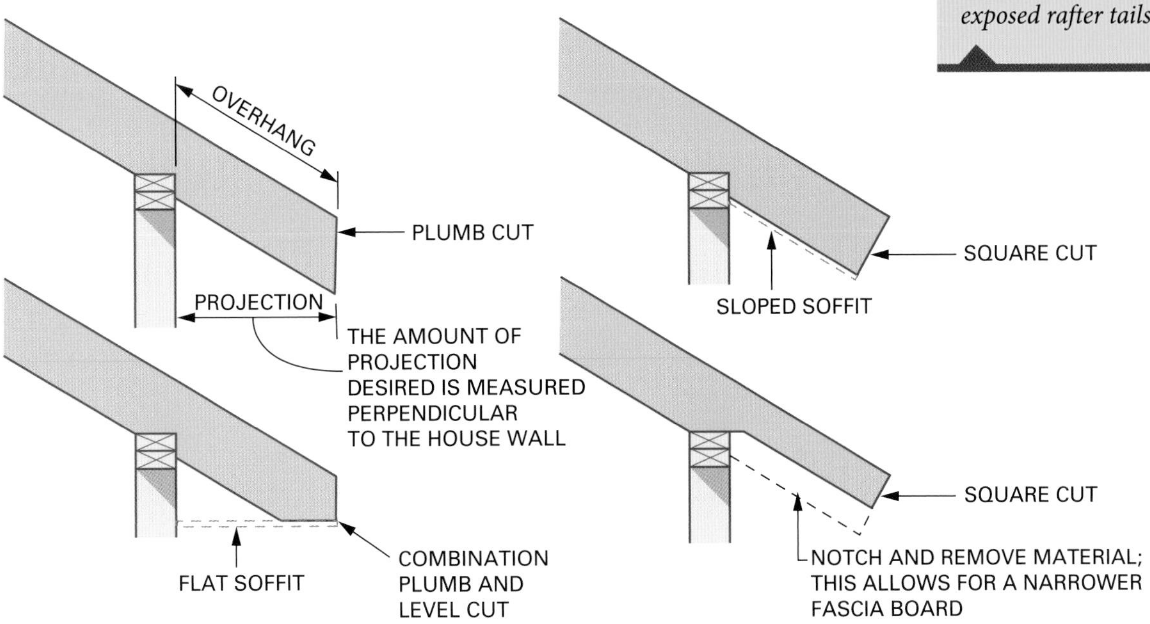
OVERHANG
PLUMB CUT
PROJECTION
THE AMOUNT OF PROJECTION DESIRED IS MEASURED PERPENDICULAR TO THE HOUSE WALL
FLAT SOFFIT
COMBINATION PLUMB AND LEVEL CUT
SQUARE CUT
SLOPED SOFFIT
SQUARE CUT
NOTCH AND REMOVE MATERIAL; THIS ALLOWS FOR A NARROWER FASCIA BOARD

Figure 6-24 Various tail cut styles

Framing the Transition from Wall to Roof

The following are two methods of framing the transition from the wall to the roof:

Method 1 Placing of the rafters on the top of the wall plate (Figure 6-25). Because it is quicker and uses less materials, this is the most common option. However, a disadvantage is that when building a house that has a steep roof and a long (2′ or so) overhang, the lower edge of the fascia may partially block the sunlight and/or view from the windows. Furthermore, this method will cut down on usable floor space in the attic area.

Method 2 As this method shows (Figure 6-26), the entire roof structure is built on top of a box header, not on the wall plate. This lifts the soffit/fascia further above the windows and creates more attic floor space. However, it does increase the labor and materials used, and adds to the overall height of the building.

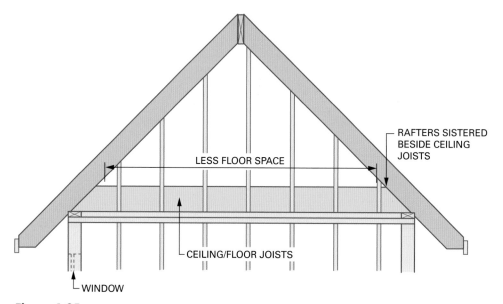

Figure 6-25 Rafters framed on top of the wall plate

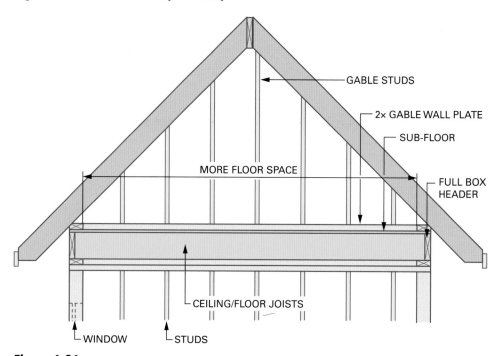

Figure 6-26 Rafters framed on top of a box header

ASSEMBLY OF A GABLE ROOF

The wall plates are first laid out to locate the rafters. The ridgeboard will be laid out with the same OC spacing with the exception of the end. The ridge layout will be offset so that the desired amount overhangs the gable end. This overhanging portion of the ridge will help to support the fly rafters (Figure 6-3).

When building a simple gable roof, all rafters are the same; therefore, only one rafter needs to be laid out; this rafter is then used as a pattern to trace and cut all of the others. Figure 6-27 shows one suggested method for erecting a gable roof.

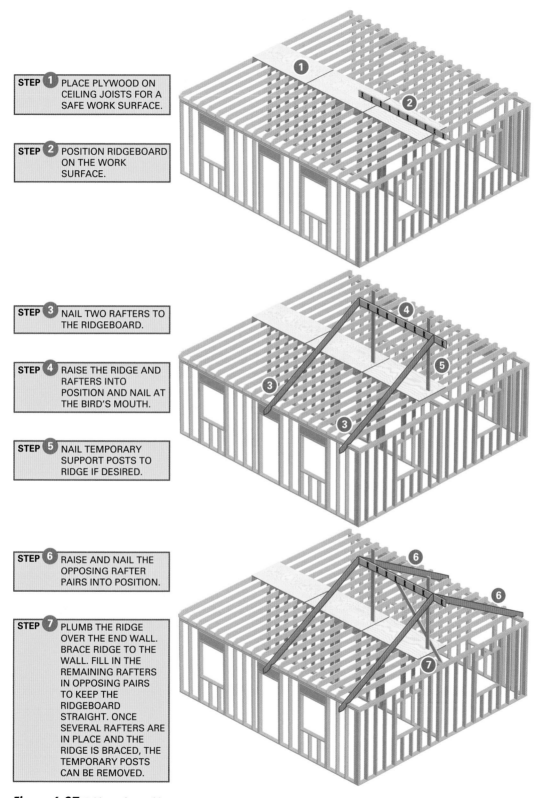

STEP 1 PLACE PLYWOOD ON CEILING JOISTS FOR A SAFE WORK SURFACE.

STEP 2 POSITION RIDGEBOARD ON THE WORK SURFACE.

STEP 3 NAIL TWO RAFTERS TO THE RIDGEBOARD.

STEP 4 RAISE THE RIDGE AND RAFTERS INTO POSITION AND NAIL AT THE BIRD'S MOUTH.

STEP 5 NAIL TEMPORARY SUPPORT POSTS TO RIDGE IF DESIRED.

STEP 6 RAISE AND NAIL THE OPPOSING RAFTER PAIRS INTO POSITION.

STEP 7 PLUMB THE RIDGE OVER THE END WALL. BRACE RIDGE TO THE WALL. FILL IN THE REMAINING RAFTERS IN OPPOSING PAIRS TO KEEP THE RIDGEBOARD STRAIGHT. ONCE SEVERAL RAFTERS ARE IN PLACE AND THE RIDGE IS BRACED, THE TEMPORARY POSTS CAN BE REMOVED.

Figure 6-27 Gable roof assembly

For an alternate method of assembly, see "Calculating Ridge-Post Height," later in this chapter.

Lookouts

The portion of the roof that overhangs the gable end is called the *rake edge*. The rafters overhanging the gable end are often called *fly rafters, rake rafters, or barge rafters*. Fly rafters are generally not as wide as the other rafters and may need extra support. This can be given in the form of cantilevered supports called *lookouts* (see Figure 6-28). The tops of lookouts are flush with the top of the rafters and can also support the roof sheathing.

Trimming Rafter Tails

Occasionally rafter tails need to be trimmed before applying the fascia board. This can be necessary due to a change in the desired projection, or to repair a previous error. The following is a method (Figure 6-29) to accurately trim the tails so that all align.

Fly rafters are the same length as common rafters but generally are not as wide and do not have bird's mouth cuts. Common rafters may be 2 × 10 or larger and fly rafters only 2 × 6 or 2 × 4, thus needing lookouts for strength.

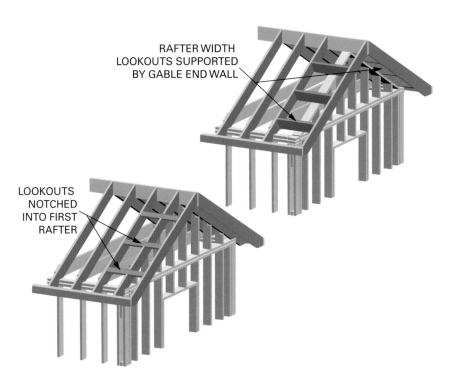

RAFTER WIDTH LOOKOUTS SUPPORTED BY GABLE END WALL

LOOKOUTS NOTCHED INTO FIRST RAFTER

Figure 6-28 Two style of lookouts can support the rake overhang

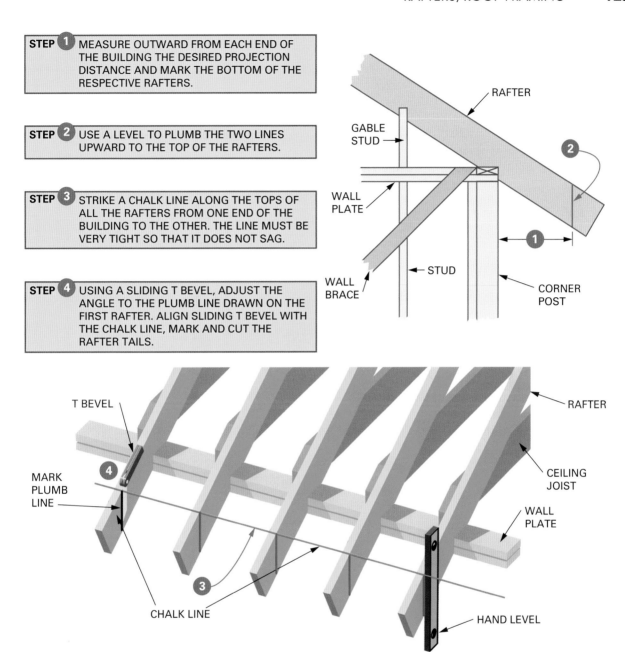

STEP 1 MEASURE OUTWARD FROM EACH END OF THE BUILDING THE DESIRED PROJECTION DISTANCE AND MARK THE BOTTOM OF THE RESPECTIVE RAFTERS.

STEP 2 USE A LEVEL TO PLUMB THE TWO LINES UPWARD TO THE TOP OF THE RAFTERS.

STEP 3 STRIKE A CHALK LINE ALONG THE TOPS OF ALL THE RAFTERS FROM ONE END OF THE BUILDING TO THE OTHER. THE LINE MUST BE VERY TIGHT SO THAT IT DOES NOT SAG.

STEP 4 USING A SLIDING T BEVEL, ADJUST THE ANGLE TO THE PLUMB LINE DRAWN ON THE FIRST RAFTER. ALIGN SLIDING T BEVEL WITH THE CHALK LINE, MARK AND CUT THE RAFTER TAILS.

Figure 6-29 Trimming rafter tails

SPECIAL CASES FOR COMMON RAFTERS

Salt Box

A salt box house is not symmetrical when viewed from the gable end (see Figure 6-30). The ridge is often not centered and one wall is higher than the other. Furthermore, each side of the roof may or may not be sloped the same; therefore, two different sets of common rafters must be calculated and cut.

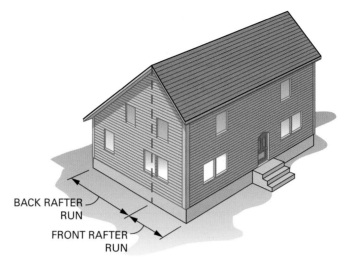

Figure 6-30 A salt box house has two different sets of common rafters

Note the two different rafter runs. To calculate the lengths, use the run and unit rise associated with each rafter. Use the formula: unit length × run = line length.

Shed Roof

A shed roof slopes in only one direction. In Figure 6-31, the building has two separate shed roofs.

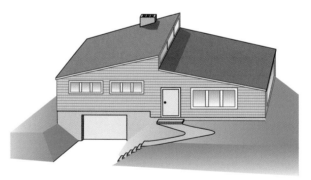

Figure 6-31 A building with shed roofs butting against each other

An attached shed roof rafter is constructed with common rafters. The difference in the layout is this: when laying out the ridge plumb cut, the shortened ridge line must be marked by subtracting the *full width of the ridge* (not half the width as with the gable roof). For example, calculate the line length. If the run is 20′, and the unit rise is 3″ (unit length = 12.37) and the ridge is 2 × material (1½″ wide):

$$\text{Line length} = 12.37 \times 20' = 247\tfrac{3}{8}''$$

ATTACHED SHED ROOF

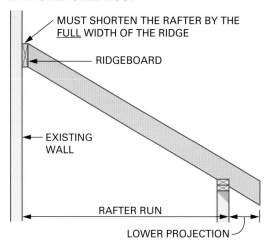

MUST SHORTEN THE RAFTER BY THE <u>FULL</u> WIDTH OF THE RIDGE

RIDGEBOARD

EXISTING WALL

RAFTER RUN

LOWER PROJECTION

FREESTANDING SHED ROOF

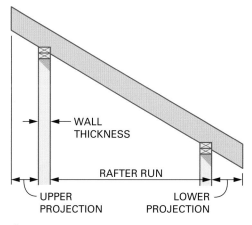

WALL THICKNESS

RAFTER RUN

UPPER PROJECTION

LOWER PROJECTION

Figure 6-32 Attached and freestanding shed roofs

During the layout, make sure to shorten the ridge plumb line by 1½″ and add the desired projection.

Freestanding shed roof rafters have two bird's mouth cuts that need to be considered. See Figure 6-32. To ensure the rafter slopes as desired, make both bird's mouth cuts to the same depth.

The freestanding shed roof rafters are calculated using the same method as above, but make sure to add for the desired projection *at each end of the rafter*. Follow Figure 6-33 for an example layout.

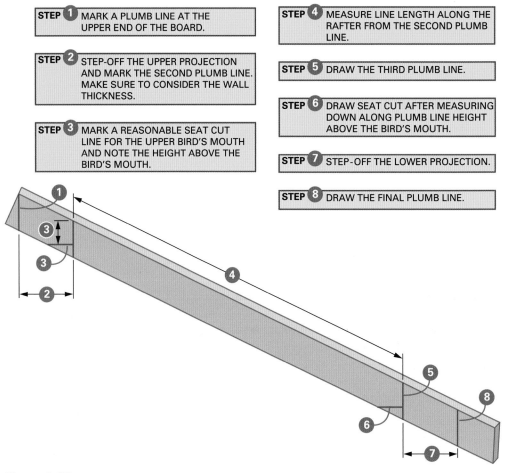

STEP **1**	MARK A PLUMB LINE AT THE UPPER END OF THE BOARD.
STEP **2**	STEP-OFF THE UPPER PROJECTION AND MARK THE SECOND PLUMB LINE. MAKE SURE TO CONSIDER THE WALL THICKNESS.
STEP **3**	MARK A REASONABLE SEAT CUT LINE FOR THE UPPER BIRD'S MOUTH AND NOTE THE HEIGHT ABOVE THE BIRD'S MOUTH.

STEP **4**	MEASURE LINE LENGTH ALONG THE RAFTER FROM THE SECOND PLUMB LINE.
STEP **5**	DRAW THE THIRD PLUMB LINE.
STEP **6**	DRAW SEAT CUT AFTER MEASURING DOWN ALONG PLUMB LINE HEIGHT ABOVE THE BIRD'S MOUTH.
STEP **7**	STEP-OFF THE LOWER PROJECTION.
STEP **8**	DRAW THE FINAL PLUMB LINE.

Figure 6-33 Laying out a freestanding shed rafter (overhanging each end)

Gambrel Roof

A gambrel roof is a symmetrical dual-pitched roof where each side contains two differently sloping common rafters. A "true" gambrel roof will have its rafter intersection points touch the sides of an imaginary circle (see Figure 6-34).

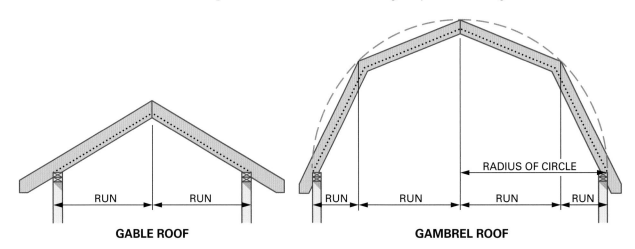

GABLE ROOF   **GAMBREL ROOF**

Figure 6-34a The gambrel has two differently sloped roof surfaces on each side

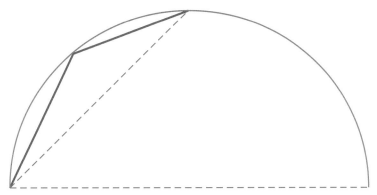

Figure 6-34b The gambrel roof allows for more space in the upper level than a gable roof

If the points of the gambrel roof do not touch the circle, this does not necessarily mean the roof is not structurally sound. There are different ways of designing and building gambrel roofs.

Gambrel roofs can be designed/built in different ways; using a circle to aid in the design is just one method.

Gambrel roofs are often seen on large barns. One reason is, obtaining rafters long/strong enough to span these large buildings was very difficult. Yet when a Gambrel roof is built, the sloping lower rafter allows for a shorter upper rafter and has the added benefit of a large upper floor storage area. This style was also adopted for houses to allow for more headroom/living space on the upper floor.

Gambrel Calculations To determine the rafter lengths and roof slopes, study Figure 6-35 and note the right triangles formed when breaking the roof into imaginary sections.

When designing or building a "true" gambrel roof, very little information is needed. The following information is known:

- Span of building = diameter of circle
- Half of the building width = radius of circle
- The ceiling height, in this case, is 8′

The intersection of R_1 and R_2 may not always be at 8′.

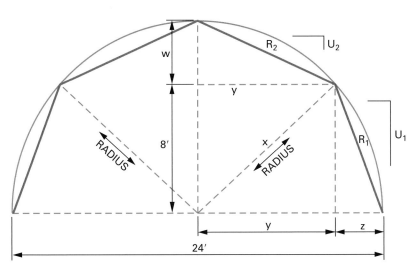

Figure 6-35 A true gambrel is built within an imaginary circle.

The following information can be inferred or calculated:

- R_1 = rafter 1, R_2 = rafter 2
- U_1 = unit rise for R_1, U_2 = unit rise for R_2
- y = run of R_2, z = run of R_1

$$x = \text{the radius} = \tfrac{1}{2} \text{ the } 24' \text{ diameter; therefore, } x = 12'$$
$$8' + w = 12 \text{ (which is the radius); therefore, } w = 4'$$

Substitute in the values from Figure 6-35 and using the Pythagorean Theorem solve.

$$8^2 + y^2 = 12^2$$
$$y^2 = 12^2 - 8^2 = 144 - 64 = 80$$
$$y = \sqrt{80} = 8.944'$$
$$y + z = 12'; \text{ therefore, } z = 3.056'$$

Rafters R_1 and R_2 can be found using the Pythagorean Theorem. Each rafter's line length is represented by the hypotenuse of the right triangle. By substituting known values into the formula:

$$8^2 + 3.056^2 = R_1^2$$
$$64 + 9.339 = R_1^2$$
$$R_1^2 = 73.339$$
$$R_1 = \sqrt{73.339} = 8.564' = 102\tfrac{3}{4}$$

Similarly,

$$4^2 + 8.944^2 = R_2^2$$
$$16 + 79.995 = R_2^2 = 95.995$$
$$R_2 = \sqrt{95.995} = 9.798' = 117\tfrac{9}{16}''$$

To find U_1, change the rise of R_1 to inches ($8 \times 12 = 96''$); the run of R_1 (z) is already in feet ($z = 3.056'$). To solve:

$$U_1 = 96 \div 3.056 = 31.41''$$

A steep roof! It slopes more than $31''$ per foot of run. Similarly,

$$U_2 = 48 \div 8.944' = 5.367''$$

Gambrel Framing Details

A common method of framing a gambrel roof utilizes a knee wall.

If framed with a knee wall (as Figure 6-36a illustrates), make sure floor joists are sized adequately to carry not only the floor load but also the roof load that will be transferred downward to the joists.

Gambrel roofs can be constructed without the use of a knee wall. Figure 6-36b illustrates the rafters fastened to a horizontal member called a purlin. The rafter typically has a notch to accommodate this method. Rafters can also be joined with metal or plywood gusset plates. There are many variations when constructing gambrel roofs.

Use the Pythagorean Theorem to solve for **y**. *The Pythagorean Theorem states* $a^2 + b^2 = c^2$; a $(8')$ *is one leg of the triangle,* b (y) *the other leg, and* c $(12')$ *the hypotenuse. The formula can be re-written to solve for* a *(or* b) *by stating* $a^2 = c^2 - b^2$ *or* $b^2 = c^2 - a^2$.

When calculating common rafters, remember, unit rise $\times$ *run = total rise (the unit rise is represented in inches, the run in feet, and total rise in inches). Therefore,* **Total Rise $\div$ Run = Unit Rise.** *Use this formula to calculate the values of* U_1 *(Unit Rise 1) and* U_2 *(Unit Rise 2).*

Some plans include the unit rises; in that case, use the values of "y" and "z" in the diagram as the run for each rafter, then use unit length $\times$ run = line length.

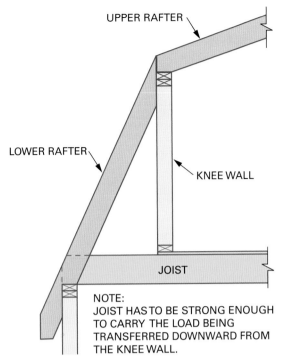

UPPER RAFTER

LOWER RAFTER

KNEE WALL

JOIST

NOTE:
JOIST HAS TO BE STRONG ENOUGH
TO CARRY THE LOAD BEING
TRANSFERRED DOWNWARD FROM
THE KNEE WALL.

Figure 6-36a Gambrel roof rafters may intersect each other
at a knee wall

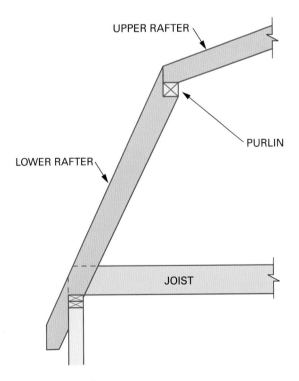

UPPER RAFTER

PURLIN

LOWER RAFTER

JOIST

Figure 6-36b Gambrel rafters may intersect each other at a purlin

Gambrel Rafter Layout

The upper rafter is a common rafter except it does not have a tail. The upper rafter tail will be cut at the seat plumb line. The lower rafter is very steeply sloped and also a common rafter. With this method of framing, the lower rafter does not get shortened at the tip (Figure 6-37).

UPPER-SLOPE RAFTER

STEP 1 LAY OUT PLUMB LINE.

STEP 2 SHORTEN ½ THICKNESS OF RIDGE.

STEP 3 LAY OUT LINE LENGTH.

STEP 4 MARK PLUMB LINE.

STEP 5 DRAW LEVEL LINE THE WIDTH OF KNEE WALL PLATE.

LOWER-SLOPE RAFTER

STEP 1 LAY OUT PLUMB LINE– NO SHORTENING.

STEP 2 LAY OUT LINE LENGTH.

STEP 3 DRAW PLUMB LINE.

STEP 4 LAY OUT LEVEL LINE OF SEAT CUT. LEAVE ON MINIMUM OF ⅔ WIDTH OF RAFTER STOCK.

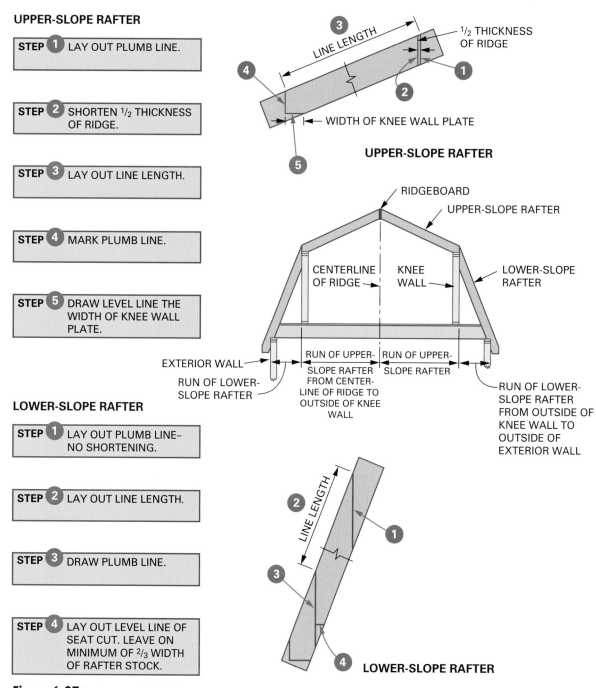

Figure 6-37 Laying out gambrel rafters

GABLE STUDS

Gable end studs should have the same OC spacing as the wall studs, and they should be positioned directly above the wall studs.

Determining Gable Stud Length

- **Measurement Method**—Studs can be individually measured as in Figure 6-38. This method is time consuming.

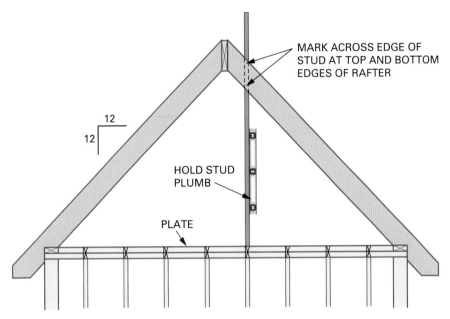

MARK ACROSS EDGE OF STUD AT TOP AND BOTTOM EDGES OF RAFTER

HOLD STUD PLUMB

PLATE

A METHOD TO FIND THE LENGTH AND ANGLE OF CUT IS TO STAND THE STUD PLUMB AND MARK IT

Figure 6-38 Cut-and-fit method of finding the length of a gable stud

- **Common Difference Method**—Stud lengths can be calculated using their common difference in length.

If the stud spacing is consistent, such as 16 OC, and the rafter slope is constant (not curved), then the height difference between one gable stud and the next is predictable (Figure 6-39). This is called a common difference (CD). In order to calculate CD for a given OC spacing and given roof slope, divide the OC spacing by 12″ (this changes the OC spacing to feet), then multiply by the unit rise.

The formula is **CD = (OC spacing ÷ 12) × unit rise**

Example 1: If the roof slope is 4 on 12 (unit rise of 4) and the OC spacing is 16″, what is the CD?

$$CD = (16 \div 12) \times 4$$
$$CD = 1.333 \times 4 = 5.33″ = 5\tfrac{5}{16}″$$

This means each OC gable stud is $5\tfrac{5}{16}″$ longer (or shorter) than the next. Once the first gable stud has been measured or calculated, simply add or subtract $5\tfrac{5}{16}″$ to find the next stud length.

> The OC spacing can be thought of as the run.

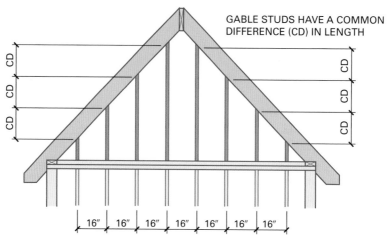

GABLE STUDS HAVE A COMMON
DIFFERENCE (CD) IN LENGTH

Figure 6-39 Equally spaced gable studs have a common difference in length

Example 2: If the roof slope is 7 on 12 and the OC spacing is 24″, what is the CD?

$$CD = (24 \div 12) \times 7$$
$$CD = 2 \times 7 = 14''$$

The smallest gable stud is often measured to insure alignment with studs in the wall below. The next stud length is the first stud plus the CD. After that, repeatedly add the CD to each previous stud's length to find the next.

Cutting Gable Studs

- **Method 1**—Gable studs can be beveled across the *entire width* of the board and then nailed to a plate that is attached either to the bottom of a rafter or to the lookouts (see Figure 6-28, right-side diagram).

- **Method 2**—Studs can be notched around the rafter as in Figure 6-40. This method is more time consuming, but the sloped top plate is unnecessary.

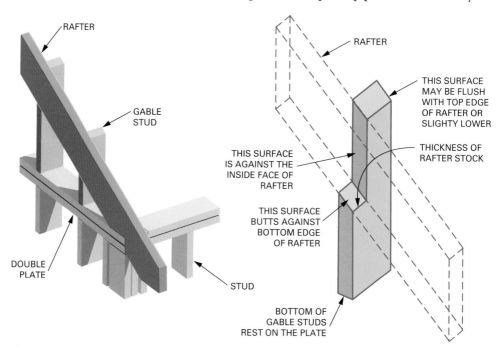

Figure 6-40 Gable stud notching method

THE HIP ROOF

A simple hip roof is found on a four-sided square or rectangular building. A complex hip roof, also known as an intersecting roof, has a differently shaped footprint that requires more roof framing details. Figure 6-41 illustrates this. Note the component names and locations.

Hipped roofs have some advantages and some disadvantages when compared to gable roofs. Advantages: They are very strong under a wind load and deflect wind better, they have level eaves providing consistent shade around the entire house, they have added protection from weather due to no high gables, and there are no high gables to paint or maintain. Disadvantages: It is more difficult to frame, there is not as much room in the attic space, and it is harder to access throughout.

Hip Component Glossary

For the following terms, see Figure 6-41.

Broken hip—A short hip rafter connecting ridges from different elevations (see Figure 6-75).

Hip jack rafter—A type of common rafter extending from the outside wall to the hip rafter.

Hip rafter—Extends diagonally (45°) from an *outside* corner of the building to the ridge.

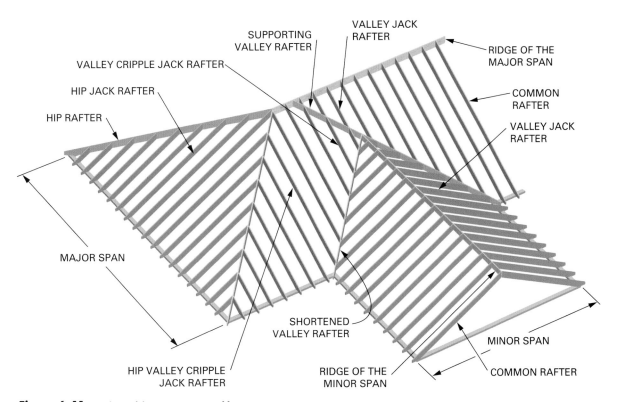

Figure 6-41 Members of the intersecting roof frame

Hip/valley cripple jack rafter—A type of common rafter that extends from a valley rafter to a hip rafter. These rafters touch neither the ridge nor the outside wall.

Shortened valley rafter—Extends diagonally (45°) from an inside corner of the building to the supporting valley rafter.

Valley cripple jack rafter—A type of common rafter that extends from a shortened valley rafter to a supporting valley rafter. These rafters touch neither the ridge nor the outside wall.

Valley jack rafter—A type of common rafter that extends from the valley rafter to the ridge.

Valley rafter—Extends diagonally (45°) from an *inside* corner of the building to the ridge. A supporting valley rafter supports a shortened valley rafter.

Hip and Valley Rafter Calculations

The calculation process to find the line lengths of hip and valley rafters is identical and layout is only slightly different.

The same formula used to calculate the line length of a common rafter is used for hip and valley rafters. **Unit length × run = line length**. However, the unit length used *must be the hip unit length*. The table of unit lengths displayed earlier is shown again below.

Notice that hip/valley unit lengths are longer than the corresponding common unit lengths—this is because a hip rafter has to travel 16.97″ to rise the same amount a common rafter rises in 12″ (Figure 6-42 and 6-43).

For every common unit of run, there is a hip unit of run. If a building has a span of 16′, its run is 8′. Therefore, there are 8 units of run for the common rafter *and* for the hip rafter. Figure 6-44 displays this concept another way.

Example: Calculate the hip unit length for a ⁵⁄₁₂ roof slope (unit rise = 5). See Figure 6-45.

To calculate a hip unit length, use the Pythagorean Theorem ($a^2 + b^2 = c^2$). The hip unit run is 16.97 and in this example, the unit rise is 5.

$$16.97^2 + 5^2 = c^2$$
$$287.981 + 25 = c^2$$
$$312.981 = c^2$$
$$c = \sqrt{312.981} = \textbf{17.691″}$$

This corresponds to the table of unit lengths shown earlier (Table 6-2). Some of the commonly used hip unit lengths can also be found on a rafter square (see Figure 6-72).

If a building has a run of 8′ and a unit rise of 5, then to calculate the hip line length, use the formula: (hip) **unit length × run = line length,** or 17.69 × 8 = 141.52″.

Because hip and valley rafters are oriented at the same angle, they are similar lengths. They use the same unit run (16.97) and therefore have identical unit lengths.

*To avoid any confusion, the roof slope on a plan is always based on the common rafters. For example, 5 on 12, **NOT** 5 on 16.97.*

*Calculating the total rise uses the same method for hip and common rafters. Use the formula **unit rise × run = total rise**. Because there are the same number of units of run for a hip and for a common rafter, the total rise will be the same. Remember, however, that the actual lengths of the unit runs are different for the hip (16.97″) and the common (12″).*

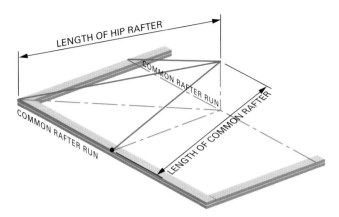

Figure 6-42 A hip rafter travels further than a common to rise an equal amount

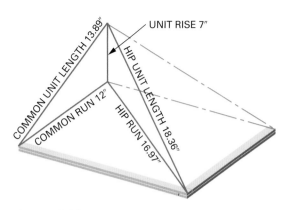

Figure 6-43 Relationship between hip and common unit triangles. This example shows a unit rise of 7″

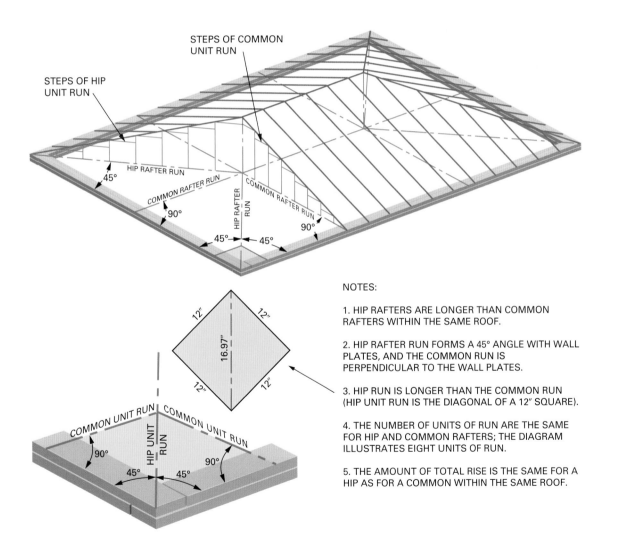

NOTES:

1. HIP RAFTERS ARE LONGER THAN COMMON RAFTERS WITHIN THE SAME ROOF.

2. HIP RAFTER RUN FORMS A 45° ANGLE WITH WALL PLATES, AND THE COMMON RUN IS PERPENDICULAR TO THE WALL PLATES.

3. HIP RUN IS LONGER THAN THE COMMON RUN (HIP UNIT RUN IS THE DIAGONAL OF A 12″ SQUARE).

4. THE NUMBER OF UNITS OF RUN ARE THE SAME FOR HIP AND COMMON RAFTERS; THE DIAGRAM ILLUSTRATES EIGHT UNITS OF RUN.

5. THE AMOUNT OF TOTAL RISE IS THE SAME FOR A HIP AS FOR A COMMON WITHIN THE SAME ROOF.

Figure 6-44 Comparing the hip and common rafters

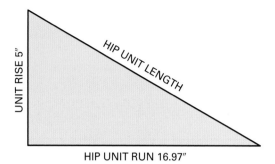

Figure 6-45 Calculating hip unit length

Table 6-2 Table of unit lengths

Unit Rise	Common Rafter Unit Length	Hip/Valley Rafter Unit Length
3	12.37	17.23
4	12.65	17.44
5	13	17.69
6	13.42	18
7	13.89	18.36
8	14.42	18.76
9	15	19.21
10	15.62	19.70
11	16.28	20.22
12	16.97	20.78

Calculating a Hip Roof Ridge Length

To make the hip roof assembly process efficient (Figure 6-70), the length of the ridgeboard should be calculated. First, however, one must decide on which of two framing methods to use (see Figure 6-46). Depending on the method chosen, the length of the ridge will be slightly different.

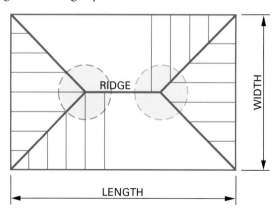

Figure 6-46 Hip framing details at the ridge

Note the different ways that rafters can meet at the end of the ridge. The green area, on the left of Figure 6-46, shows a common rafter joining the *end* of the ridge. The orange area, on the right of Figure 6-46, shows only the hip rafters touching the end of the ridge. Figure 6-47 shows an enlarged detail.

To calculate the ridge length, subtract the width of the building from the length, and using Figure 6-47, decide how much more must be added to each end of the ridge depending on the framing detail. For example, if the building in Figure 6-46 is 20′ long and 10′ wide, and the framing material is 2× (meaning 1½″ thick), the ridge length will be 20′ minus 10′ = 10′ plus ¾″ for the detail in the green area, plus ¾″ + 1¹⁄₁₆″ (1¹³⁄₁₆″) for the detail in the orange area. This will give an overall width of 10′ + ¾″ + ¾″ + 1¹⁄₁₆″ = 10′ 2⁹⁄₁₆″. **Note:** For the purpose of this example, each end of the roof (see Figure 6-46) is framed differently.

TIP

Determining the method of framing at the ridge may be based on either personal preference or on where the layout marks fall. This can be determined by following the procedure in Figure 6-69.

The measurements shown in Figure 6-47 correspond to framing members that are 1½″ (2×) thick. If the ridge and/or rafters are a different thickness than shown, adjustments may be necessary.

It may be necessary to align the square along the top edge of the board as shown (see Figure 6-48) or along the bottom edge (see Figure 6-13). Either method will produce the same result but one method makes it easier to draw a continuous plumb line. Positioning of the square depends on the depth of the rafter and the slope of the roof.

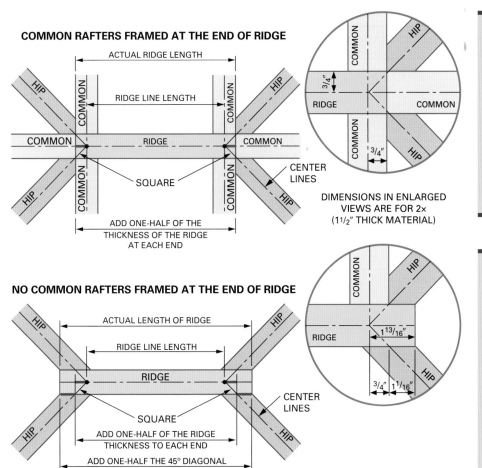

Figure 6-47 Determining the actual length of the hip roof ridge

LAYOUT OF HIP RAFTERS

For this section, all lumber in the diagrams is 2× (meaning 1½″ thick) unless noted otherwise.

Hip rafter layout is more complex than common rafter layout. There are several additional steps that must be drawn in order to accurately produce a hip rafter. This procedure is broken down into several parts: the tip, cheek cuts, line length, the seat, and the tail. For layout purposes, 17″ is used as the hip run; for calculation purposes (accuracy), 16.97 is used.

The Tip of the Hip Rafter

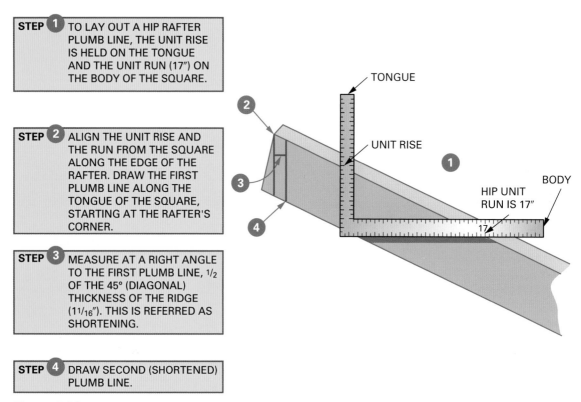

STEP 1	TO LAY OUT A HIP RAFTER PLUMB LINE, THE UNIT RISE IS HELD ON THE TONGUE AND THE UNIT RUN (17″) ON THE BODY OF THE SQUARE.
STEP 2	ALIGN THE UNIT RISE AND THE RUN FROM THE SQUARE ALONG THE EDGE OF THE RAFTER. DRAW THE FIRST PLUMB LINE ALONG THE TONGUE OF THE SQUARE, STARTING AT THE RAFTER'S CORNER.
STEP 3	MEASURE AT A RIGHT ANGLE TO THE FIRST PLUMB LINE, ½ OF THE 45° (DIAGONAL) THICKNESS OF THE RIDGE (1¹/₁₆″). THIS IS REFERRED AS SHORTENING.
STEP 4	DRAW SECOND (SHORTENED) PLUMB LINE.

Figure 6-48 Marking a hip rafter plumb line

Cheek Cut Layout for a Hip Rafter

Remember, there are two different methods of framing at the ridge, thus two ways to lay out the cheek cuts. If a single cheek cut is needed, stop at Step 5 (Figure 6-49); if a double cheek is needed, continue through Steps 6–8.

CHEEK CUTS USING MEASUREMENT METHOD

SINGLE CHEEK LINE

STEP 1 SQUARE A LINE ACROSS THE TOP EDGE FROM THE SECOND PLUMB LINE.

STEP 2 MARK THE CENTER OF THE TOP EDGE.

STEP 3 FROM THE SECOND LINE, MEASURE AT A RIGHT ANGLE $1/2$ THE THICKNESS OF THE HIP RAFTER. IT IS $3/4''$ FOR DIMENSION LUMBER.

STEP 4 DRAW THIRD PLUMB LINE. THIS LINE IS THE CUT LINE.

STEP 5 DRAW A DIAGONAL LINE ACROSS THE TOP FROM THE THIRD PLUMB LINE THROUGH THE CENTERLINE. THIS SHORTENS THE RAFTER THE SECOND TIME AND WILL REPRESENT THE ANGLE OF THE CUT.

NOTE:
THE ANGLE OF THE DIAGONAL LINE CAN BE MEASURED WITH A SPEED SQUARE OR A PROTRACTOR; THIS IS THE BEVEL ANGLE THAT THE CHEEKS WILL BE CUT.

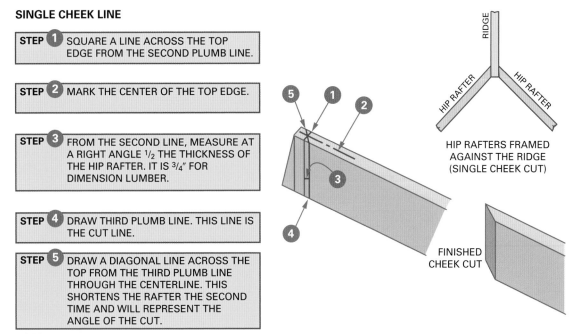

HIP RAFTERS FRAMED AGAINST THE RIDGE (SINGLE CHEEK CUT)

FINISHED CHEEK CUT

DOUBLE CHEEK LINES

STEP 6 SQUARE ACROSS FROM THE THIRD PLUMB LINE.

STEP 7 DRAW A DIAGONAL LINE FROM SQUARED LINE THROUGH THE CENTERLINE.

STEP 8 A FOURTH PLUMB LINE MAY BE DRAWN AS A CUT LINE.

NOTE:
TO CUT THE DOUBLE CHEEKS, FIRST CUT LINE #8 WITH THE SAW BEVELED IN THE DIRECTION OF THE BLUE LINE (#7) ON THE TOP OF THE RAFTER. NEXT CUT LINE #4, WITH THE BEVEL ANGLE FOLLOWING THE BLUE LINE ON THE TOP OF THE RAFTER THAT CORRESPONDS WITH LINE #4. THIS WILL PRODUCE A POINTED TIP.

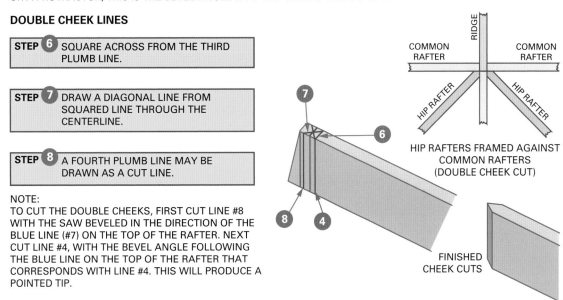

HIP RAFTERS FRAMED AGAINST COMMON RAFTERS (DOUBLE CHEEK CUT)

FINISHED CHEEK CUTS

Figure 6-49 Making cheek cuts using the measurement method

The rafter must be shortened to accommodate the ridge, and then shortened again to accommodate the cheek cut(s) of the hip rafter (see Figure 6-49). A 1½" thick framing member beveled at 45° measures 2⅛" across the face of the bevel; ½ of the beveled face is 1¹⁄₁₆".

Alternate Cheek Cut Layout There is another method of laying out the cheek cut angles, and that is by using the side-cut tables found on the rafter square. Figure 6-50 illustrates this technique.

- **Step 1**—If the slope of the roof is ⁴⁄₁₂, find the side-cut number on the 6th line down under the number **4** the rafter square.
- **Step 2**—Hold **12** on the body of the square and the number found in Step 1 on the tongue of the square; mark along *body* on the edge of the board as illustrated in Figure 6-50.
- **Step 3**—Using a speed square, measure the degree angle. Note that the angle will be greater than 45°, and the steeper the roof slope, the greater the number.
- **Step 4**—The measured angle is the bevel angle the saw is set to for making the cheek cut(s).

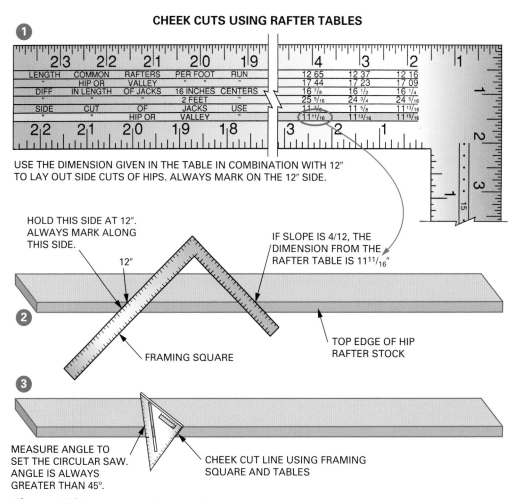

Figure 6-50 Cheek cuts using the rafter table method

Measuring Line Length

Line length can either be calculated (hip unit length × run = line length) and then measured (see Figure 6-51, top), or stepped off with a square (see Figure 6-51, bottom).

To lay out a fractional unit of run on the hip rafter, a calculation based on the common run must be made.

Example: A building has a run of 8′ 9″ and a unit rise of 7″. The first 8 units of run are stepped off by holding 7″ on the tongue of the square and 17″ on the body. To step-off the fractional amount of 9″ (which is based on the common

> *Earlier in the chapter the unit run for a hip rafter was found to be 16.97″. The number 16.97″ is used for calculations; but for layout, 17″ is close enough.*

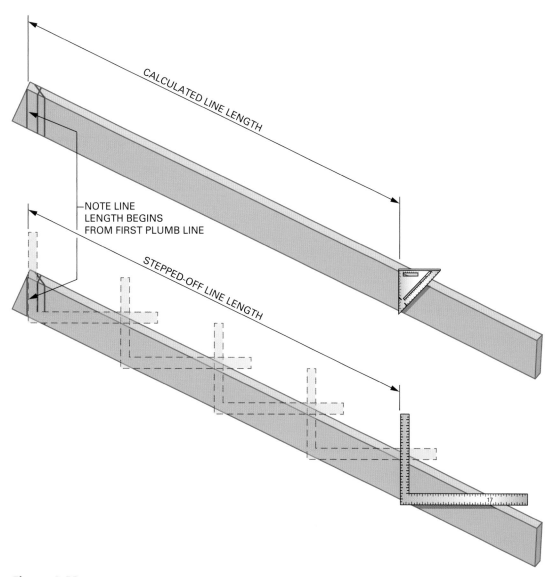

CALCULATED LINE LENGTH

NOTE LINE LENGTH BEGINS FROM FIRST PLUMB LINE

STEPPED-OFF LINE LENGTH

Figure 6-51 Line length can either be calculated and measured or stepped off. In both cases, begin at the first plumb line

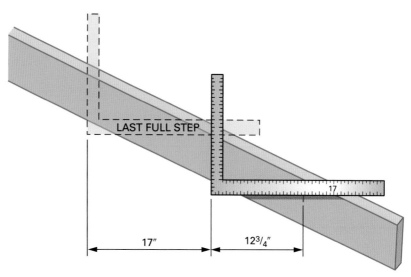

Figure 6-52 Stepping off a fractional unit of run on the hip rafter

rafter run), consider that 9″ is actually $\frac{9}{12}$ of a unit of run; $\frac{9}{12} = \frac{3}{4}$, and $\frac{3}{4}$ of 17″ = 12¾″. Therefore the extra 9″ of run for a common rafter is actually 12¾″ for a hip rafter. See Figures 6-51 and 6-52.

The Seat Cut of a Hip Rafter

Figure 6-53 illustrates the process of laying out the seat cut for a hip rafter. Because the hip rafter runs at an angle to the common rafters, it must either be lowered ("dropped") or the entire length of the hip rafter must be beveled ("backed"). Backing the rafter is time consuming, so most carpenters prefer to drop the rafter by cutting the seat a little deeper.

The Tail of the Hip Rafter

Hip rafter tails are much more complex than common tails. One difference is that the very end of the tail is cut with a double bevel; this allows the rafter to accommodate the fascia boards from two directions.

There are two methods of calculation/layout: the overhang method and the projection method.

- **Method 1**—the overhang method (see Figure 6-54).
- **Method 2**—the projection method.

  The lower-right-side graphic in Figure 6-54 represents the *projection* of two common rafter tails (15″) and the *projection* of the corresponding hip rafter tail (21³⁄₁₆″), which is the diagonal of the square. Using the Pythagorean Theorem, substitute the common projection dimensions for **a** and **b**, and then solve for **c**, which represents the *projection* of the hip rafter. For example, if the common projection is 15″, then $15^2 + 15^2 = c^2$; $450 = c^2$; $c = \sqrt{450}$; therefore c = 21³⁄₁₆″.

Common rafters are generally placed before hip rafters. Thus, the height above the bird's mouth can be measured directly from a common rafter.

Remember, any plumb line on a hip rafter is drawn by holding the unit rise on the tongue of the square and 17″ (hip run) on the body.

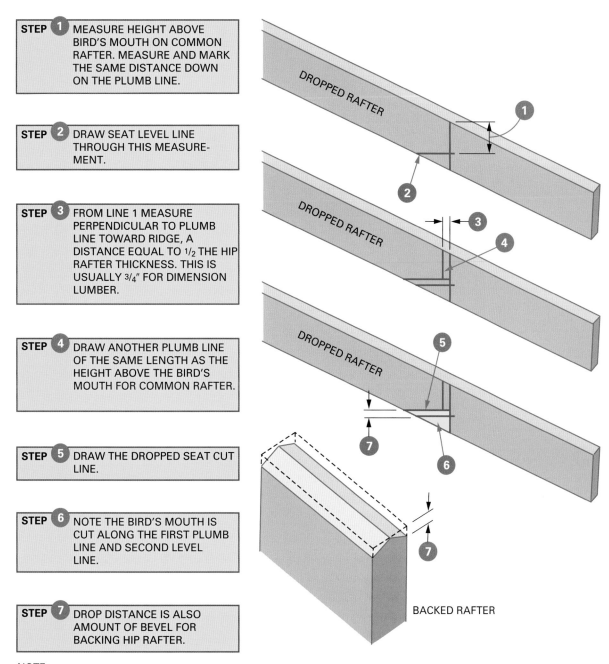

STEP 1 MEASURE HEIGHT ABOVE BIRD'S MOUTH ON COMMON RAFTER. MEASURE AND MARK THE SAME DISTANCE DOWN ON THE PLUMB LINE.

STEP 2 DRAW SEAT LEVEL LINE THROUGH THIS MEASUREMENT.

STEP 3 FROM LINE 1 MEASURE PERPENDICULAR TO PLUMB LINE TOWARD RIDGE, A DISTANCE EQUAL TO 1/2 THE HIP RAFTER THICKNESS. THIS IS USUALLY 3/4″ FOR DIMENSION LUMBER.

STEP 4 DRAW ANOTHER PLUMB LINE OF THE SAME LENGTH AS THE HEIGHT ABOVE THE BIRD'S MOUTH FOR COMMON RAFTER.

STEP 5 DRAW THE DROPPED SEAT CUT LINE.

STEP 6 NOTE THE BIRD'S MOUTH IS CUT ALONG THE FIRST PLUMB LINE AND SECOND LEVEL LINE.

STEP 7 DROP DISTANCE IS ALSO AMOUNT OF BEVEL FOR BACKING HIP RAFTER.

BACKED RAFTER

NOTE:
BACKING THE RAFTER IS NECESSARY ONLY IF THE SEAT HAS NOT BEEN DROPPED

Figure 6-53 Dropping or backing a hip rafter

If using the projection method, the rafter tail is laid out by holding the tongue of the square along the seat plumb line (the first one drawn) and measuring along the body toward the tail 21³⁄₁₆″. Mark the tail and draw a plumb line through the mark. This represents Line 5 in Figure 6-54. Continue through Steps 6–10 to complete the layout.

When using the overhang method of layout, follow Steps 1–10.

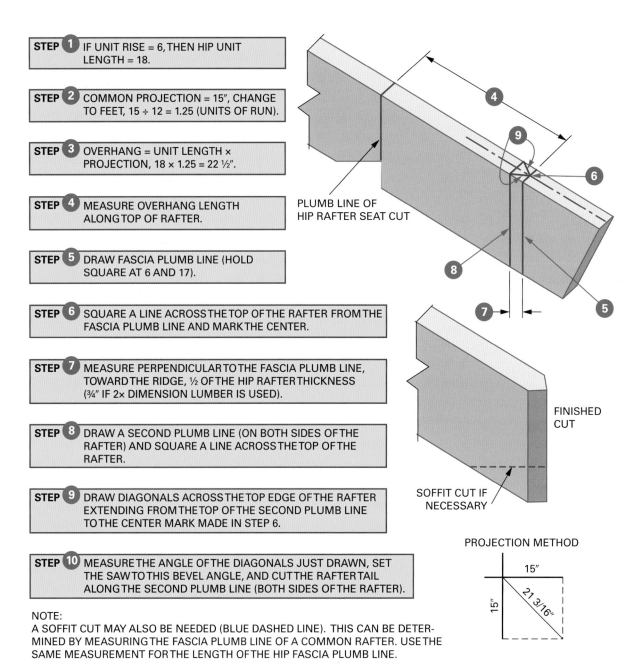

STEP 1 IF UNIT RISE = 6, THEN HIP UNIT LENGTH = 18.

STEP 2 COMMON PROJECTION = 15″, CHANGE TO FEET, 15 ÷ 12 = 1.25 (UNITS OF RUN).

STEP 3 OVERHANG = UNIT LENGTH × PROJECTION, 18 × 1.25 = 22 ½″.

STEP 4 MEASURE OVERHANG LENGTH ALONG TOP OF RAFTER.

STEP 5 DRAW FASCIA PLUMB LINE (HOLD SQUARE AT 6 AND 17).

STEP 6 SQUARE A LINE ACROSS THE TOP OF THE RAFTER FROM THE FASCIA PLUMB LINE AND MARK THE CENTER.

STEP 7 MEASURE PERPENDICULAR TO THE FASCIA PLUMB LINE, TOWARD THE RIDGE, ½ OF THE HIP RAFTER THICKNESS (¾″ IF 2× DIMENSION LUMBER IS USED).

STEP 8 DRAW A SECOND PLUMB LINE (ON BOTH SIDES OF THE RAFTER) AND SQUARE A LINE ACROSS THE TOP OF THE RAFTER.

STEP 9 DRAW DIAGONALS ACROSS THE TOP EDGE OF THE RAFTER EXTENDING FROM THE TOP OF THE SECOND PLUMB LINE TO THE CENTER MARK MADE IN STEP 6.

STEP 10 MEASURE THE ANGLE OF THE DIAGONALS JUST DRAWN, SET THE SAW TO THIS BEVEL ANGLE, AND CUT THE RAFTER TAIL ALONG THE SECOND PLUMB LINE (BOTH SIDES OF THE RAFTER).

PLUMB LINE OF HIP RAFTER SEAT CUT

FINISHED CUT

SOFFIT CUT IF NECESSARY

PROJECTION METHOD

15″

15″

21 3/16″

NOTE:
A SOFFIT CUT MAY ALSO BE NEEDED (BLUE DASHED LINE). THIS CAN BE DETERMINED BY MEASURING THE FASCIA PLUMB LINE OF A COMMON RAFTER. USE THE SAME MEASUREMENT FOR THE LENGTH OF THE HIP FASCIA PLUMB LINE.

Figure 6-54 Laying out a hip rafter tail

LAYOUT OF THE VALLEY RAFTER

In the event the valley rafter is a supporting valley rafter, the layout is nearly identical. Furthermore, line length calculations are identical to hip rafter calculations: (hip) unit length × run = line length. Remember, hip and valley unit lengths are identical.

The upper right portion of Figure 6-55 illustrates how the run of the valley rafter is measured. As before, all lumber shown in the diagrams is 2 × (1½″ thick).

Valley rafters are found in intersecting roofs. As with hip rafters, there are two ways to lay out the tip depending on the roof type and framing.

Layout of the Tip of the Valley Rafter

The valley rafter layout will be broken into three parts; the first part will focus on the tip of the rafter. Steps 1–7 in Figure 6-55 diagram a single cheek cut layout; see Steps 8–10 if double cheeks are needed.

Cheek cut angles can be either laid out as in Figure 6-55, or the side-cut method can be used as in Figure 6-50.

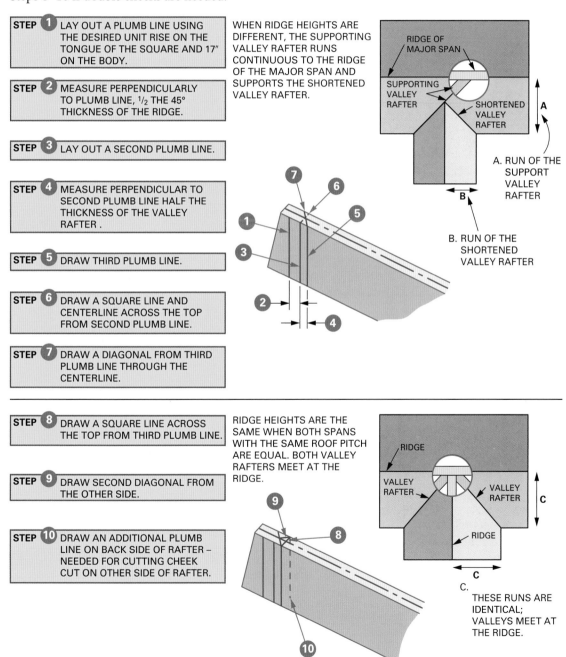

STEP 1 LAY OUT A PLUMB LINE USING THE DESIRED UNIT RISE ON THE TONGUE OF THE SQUARE AND 17″ ON THE BODY.

STEP 2 MEASURE PERPENDICULARLY TO PLUMB LINE, ½ THE 45° THICKNESS OF THE RIDGE.

STEP 3 LAY OUT A SECOND PLUMB LINE.

STEP 4 MEASURE PERPENDICULAR TO SECOND PLUMB LINE HALF THE THICKNESS OF THE VALLEY RAFTER.

STEP 5 DRAW THIRD PLUMB LINE.

STEP 6 DRAW A SQUARE LINE AND CENTERLINE ACROSS THE TOP FROM SECOND PLUMB LINE.

STEP 7 DRAW A DIAGONAL FROM THIRD PLUMB LINE THROUGH THE CENTERLINE.

STEP 8 DRAW A SQUARE LINE ACROSS THE TOP FROM THIRD PLUMB LINE.

STEP 9 DRAW SECOND DIAGONAL FROM THE OTHER SIDE.

STEP 10 DRAW AN ADDITIONAL PLUMB LINE ON BACK SIDE OF RAFTER – NEEDED FOR CUTTING CHEEK CUT ON OTHER SIDE OF RAFTER.

WHEN RIDGE HEIGHTS ARE DIFFERENT, THE SUPPORTING VALLEY RAFTER RUNS CONTINUOUS TO THE RIDGE OF THE MAJOR SPAN AND SUPPORTS THE SHORTENED VALLEY RAFTER.

A. RUN OF THE SUPPORT VALLEY RAFTER
B. RUN OF THE SHORTENED VALLEY RAFTER

RIDGE HEIGHTS ARE THE SAME WHEN BOTH SPANS WITH THE SAME ROOF PITCH ARE EQUAL. BOTH VALLEY RAFTERS MEET AT THE RIDGE.

C. THESE RUNS ARE IDENTICAL; VALLEYS MEET AT THE RIDGE.

Figure 6-55 Laying out the tip of a valley rafter

Layout of the Seat of the Valley Rafter

The second part of the valley rafter layout will focus on the seat cut. See Figure 6-56.

STEP 1 MEASURE AND MARK THE LINE LENGTH FROM THE FIRST PLUMB LINE DRAWN AT THE TIP; MARK THE SEAT PLUMB LINE (UNIT RISE ON TONGUE OF SQUARE, HIP UNIT RUN [17"] ON BODY).

STEP 2 MEASURE DOWN ALONG THE SEAT PLUMB LINE THE SAME DISTANCE AS HEIGHT ABOVE BIRD'S MOUTH ON THE COMMON RAFTER.

STEP 3 DRAW A LEVEL LINE FOR THE SEAT TO THE BOTTOM EDGE OF THE RAFTER (THIS LINE IS DRAWN SQUARE TO THE SEAT PLUMB LINE). ALSO EXTEND THIS LINE THROUGH THE OTHER SIDE OF THE PLUMB LINE (#2) AN INCH OR SO.

STEP 4 MEASURE PERPENDICULAR TO THE PLUMB LINE (#2) ALONG THE EXTENDED SEAT LINE TOWARD THE TAIL ½ OF THE THICKNESS OF THE VALLEY RAFTER (¾" IF 2× DIMENSION LUMBER IS USED).

STEP 5 DRAW A SECOND PLUMB LINE FROM THE EXTENDED SEAT LINE DOWNWARD TO THE BOTTOM EDGE OF THE RAFTER.

STEP 6 THE BIRD'S MOUTH MAY BE CUT SQUARE FROM THE SECOND PLUMB LINE, OR IF THE RAFTER TAIL IS EXPOSED, A DECORATIVE DOUBLE BEVEL CAN BE CUT.

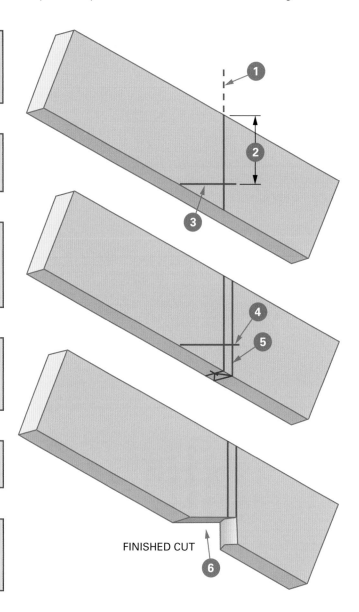

FINISHED CUT

Figure 6-56 Laying out the seat of a valley rafter

Layout of the Tail of the Valley Rafter

The third part of the valley rafter layout focuses on the tail section of the rafter.
See Figure 6-57.

STEP ①	MEASURE TAIL LENGTH FROM THE FIRST BIRD'S MOUTH PLUMB LINE.

STEP ②	DRAW PLUMB LINE (UNIT RISE ON TONGUE OF SQUARE, UNIT RUN [17"] ON BODY).

STEP ③	MEASURE PERPENDICULAR TO THE PLUMB LINE ½ THE THICKNESS OF THE VALLEY RAFTER (¾" IF USING 2× DIMENSION LUMBER).

STEP ④	SQUARE LINES ACROSS TOP OF RAFTER AS SHOWN.

STEP ⑤	DRAW DIAGONAL. NOTE, CHEEK CUT ANGLES CAN ALSO BE LAID OUT AS IN FIGURE 6-50.

NOTE:
TO SAVE TIME, A SINGLE CHEEK CUT CAN BE UTILIZED IF THE TAILS WILL BE HIDDEN WITH A SOFFIT. SEE BOTTOM OF FIGURE 6-57.

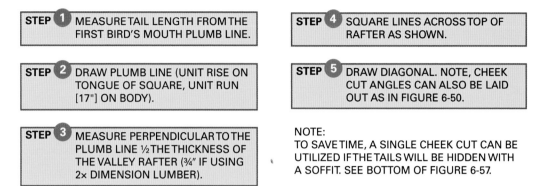

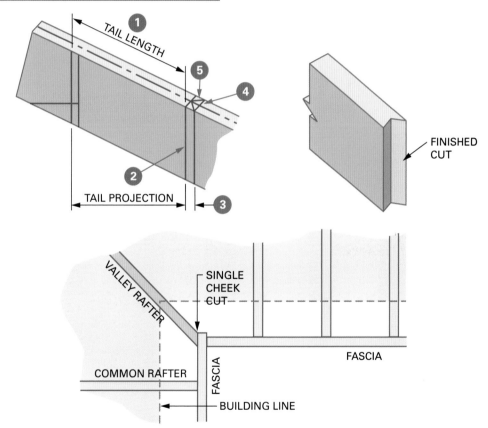

Figure 6-57 Laying out the valley rafter tail

Backing the Valley Rafter

If the valley rafter is acting as a supporting valley rafter in an intersecting roof (see Figure 6-41), then the portion between the shortened valley rafter and the ridge will have to be beveled (backed); Figure 6-58 illustrates this process.

To "back" the upper portion of the rafter, use a speed square to measure the angle of the rafter plumb cut in degrees. Set a circular saw to this angle and rip the rafter to its center so that only the portion of the supporting valley between the shortened valley and the ridge is beveled. The backing can also be accomplished with a sharp hand plane; this may be necessary if the rafter is already in place.

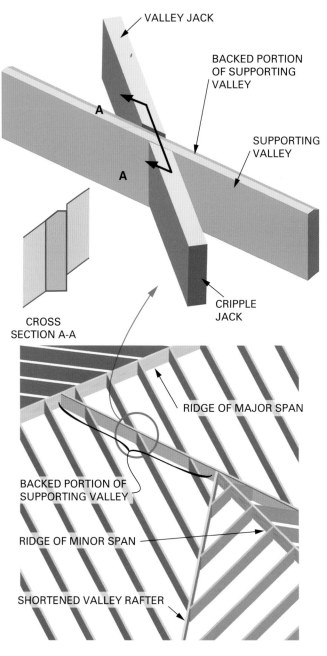

Figure 6-58 Backing the upper section of a supporting valley rafter

OTHER HIP ROOF COMPONENTS

Calculation of the Shortened Valley Rafter

The shortened valley rafter abuts the supporting valley rafter and does not connect to the ridge. Therefore, it is shorter than the supporting valley rafter.

The length of the shortened valley rafter is found by using run "A" in Figure 6-59. Use the formula: line length = hip unit length × run.

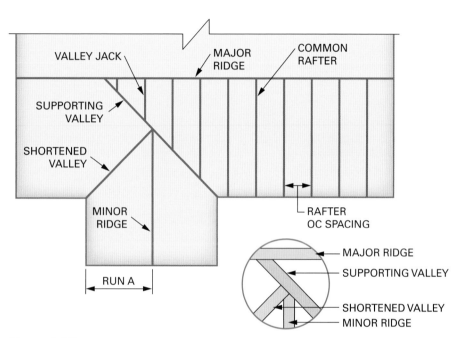

Figure 6-59 Determining the length of a shortened valley rafter

Layout of Shortened Valley Rafter

Follow the same procedures of laying out a valley rafter *except* the tip (see Figure 6-60).

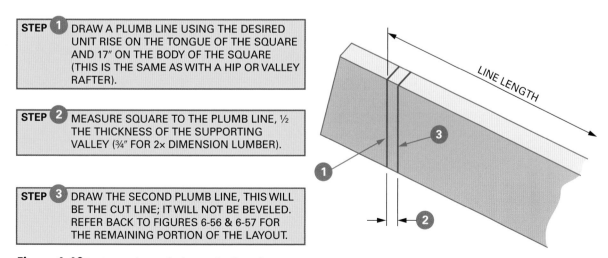

STEP 1 DRAW A PLUMB LINE USING THE DESIRED UNIT RISE ON THE TONGUE OF THE SQUARE AND 17″ ON THE BODY OF THE SQUARE (THIS IS THE SAME AS WITH A HIP OR VALLEY RAFTER).

STEP 2 MEASURE SQUARE TO THE PLUMB LINE, ½ THE THICKNESS OF THE SUPPORTING VALLEY (¾″ FOR 2× DIMENSION LUMBER).

STEP 3 DRAW THE SECOND PLUMB LINE, THIS WILL BE THE CUT LINE; IT WILL NOT BE BEVELED. REFER BACK TO FIGURES 6-56 & 6-57 FOR THE REMAINING PORTION OF THE LAYOUT.

Figure 6-60 Laying out the tip of a shortened valley rafter

Calculation of Hip Jack Rafters

Hip jack rafters are special common rafters because they are found in the same plane as the common rafters. The tail details are identical to common rafters. The run of the hip jack rafter is the horizontal distance of the jack rafter, measured directly beneath it. An easier way to determine jack run is to measure from the corner of the building, along the plate, to the center of the jack rafter; this is equivalent to the *run* (see Figure 6-61).

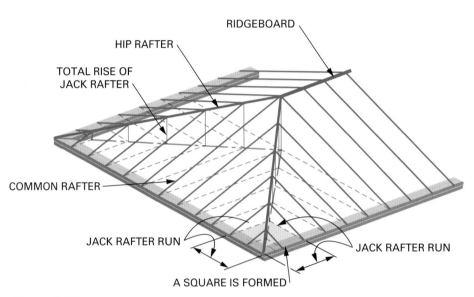

Figure 6-61 Determining jack rafter length

> Each jack is a specific length longer than the previous jack rafter. This length is dependent on the OC spacing and the slope of the roof. For example, the jack CD for a 4/12 roof and the jack CD for a 9/12 roof are different. The CD for jacks spaced 16 OC vs. 24 OC is also different.

The run of a jack rafter is equal to its distance from the corner of the building. For example, assume the center of a jack rafter is 46″ from the corner of the building. The unit rise of the roof is 4 (⁴⁄₁₂ slope). How long is the jack? Convert the run from inches to feet, $46 \div 12 = 3.833'$. If unit rise = 4″, then unit length = 12.65″. Use the formula unit length × run = line length: $12.65 \times 3.833' = 48.49'' = 48\frac{1}{2}''$.

Again, any jack rafter run can be found by measuring from the corner of the building to the center of the rafter. Furthermore, each jack rafter is a certain length longer than the previous jack rafter. This length is known as the common difference (CD) in length (Figure 6-62).

The following procedure demonstrates a method for calculating CD for jack rafters.

THE DIFFERENCE IN THE RUN OF ONE JACK RAFTER TO THE NEXT IS THE SAME AS THE OC SPACING. THEREFORE, ONCE THIS DIFFERENCE IS CALCULATED, IT CAN BE EITHER ADDED OR SUBTRACTED FROM THE PREVIOUS JACK RAFTER'S LENGTH, TO FIND THE LENGTH OF THE NEXT ONE.

EXAMPLE: IF THE UNIT RISE IS 6" AND THE RAFTER SPACING IS 16" OC, CALCULATE THE COMMON DIFFERENCE IN LENGTH FOR A JACK.

STEP **1** IF UNIT RISE = 6", THEN THE UNIT LENGTH IS 13.42 FOR A COMMON RAFTER.

STEP **2** CONVERT 16" (OC SPACING) TO FEET (UNITS OF RUN); 16 ÷ 12 = 1.333' (1⅓ UNITS OF RUN).

STEP **3** CALCULATE LINE LENGTH: (UNIT LENGTH × RUN) 13.42 × 1.333 = 17.89" = 17⅞". THEREFORE, THE COMMON DIFFERENCE FOR JACKS ON A 6/12 ROOF SPACED 16" OC IS 17⅞".

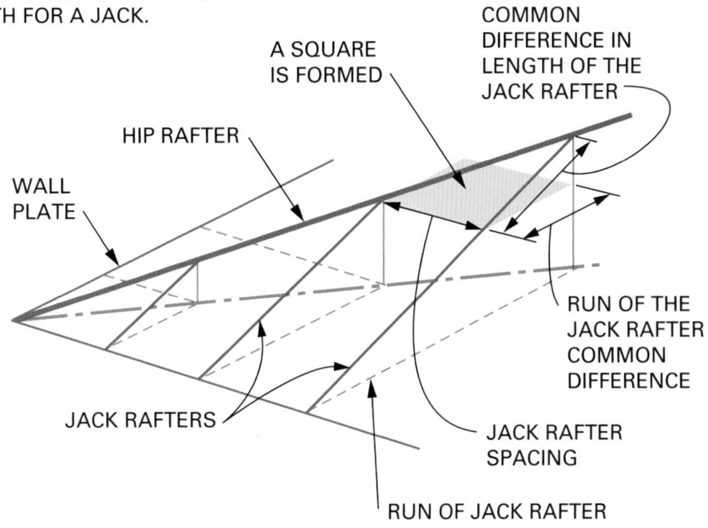

STEP **4** COMMON DIFFERENCE IS 17⅞". NOTE: THIS VALUE CAN ALSO BE FOUND ON A RAFTER SQUARE, THE THIRD LINE DOWN FROM THE TOP UNDER 6"; SEE FIGURE 6-72.

Figure 6-62 Calculating jack rafter common difference

Some CD lengths for jacks 16 OC and 24 OC are found on the rafter square (the third and fourth lines, respectively, underneath the unit rise number [see Figure 6-72]).

Layout of the Hip Jack Rafter

Hip jacks are common rafters; therefore, the seat and tail are identical to the common rafter, but the tip abuts a hip rafter at an angle (see Figure 6-63).

TIP

Carpenters often save time cutting hip jack rafters by making an extra tailpiece, including the bird's mouth (a template); they first trace the tail, then measure the line length, and finally lay-out the tip. This saves the time and potential inaccuracies from repetitive measuring and marking.

STEP 1 LAY OUT PLUMB LINE. THIS IS LAID OUT WITH THE UNIT RISE ON THE TONGUE AND 12″ (UNIT RUN) ON THE BODY OF THE SQUARE.

STEP 2 MEASURE SQUARE TO THE PLUMB LINE ½ OF THE 45° THICKNESS OF THE HIP RAFTER (1¹¹⁄₁₆″ IF 2× DIMENSION LUMBER IS USED) AND DRAW THE SHORTENED PLUMB LINE.

STEP 3 SQUARE SHORTENED PLUMB LINE ACROSS THE TOP EDGE OF THE RAFTER.

STEP 4 MEASURE AT RIGHT ANGLE FROM SHORTENED PLUMB LINE ½ THE THICKNESS OF THE JACK RAFTER STOCK AND DRAW ANOTHER PLUMB LINE.

STEP 5 DRAW CENTERLINE ALONG THE TOP OF RAFTER.

STEP 6 DRAW DIAGONAL FROM THE LAST PLUMB LINE THROUGH THE CENTERLINE. NOTE: THIS LINE MAY NEED TO BE DRAWN ON THE OPPOSITE SIDE OF THE RAFTER IF THE CUT NEEDS TO BE REVERSED.

STEP 7 THE LINE LENGTH CAN BE MEASURED FROM THE FIRST PLUMB LINE (STEP 1); THE BIRD'S MOUTH AND TAIL CAN BE DRAWN JUST LIKE THE COMMON RAFTER IN FIGURE 6-20.

NOTE:
THE CHEEK CUTS CAN BE LAID OUT USING THE SIDE-CUT METHOD (FIGURE 6-50). USE THE FIFTH LINE DOWN UNDER THE CORRESPONDING UNIT RISE TO FIND "SIDE CUT OF JACKS USE…".

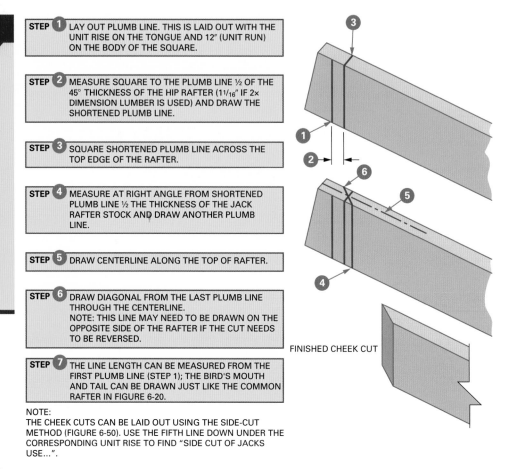

FINISHED CHEEK CUT

Figure 6-63 Laying out the tip of a hip jack rafter

Always use the common rafter unit lengths when calculating hip or valley jack rafters. Jack rafters are considered common rafters because they lie in the same plane.

Calculation of Valley Jack Rafters

Valley jack rafters are special common rafters because they are found in the same plane as the common rafters. They attach to the ridge the same way a common rafter does; however, their tails attach directly to a valley rafter. The valley jack run is found by measuring horizontally from the intersection of the ridge/valley centerlines to the center of the valley jack (see Figure 6-64).

Once the run is obtained, use the line length formula (unit length × run) to find the jack rafter length. For example, to find the line length of a valley jack if the valley jack run is 20″ (1.667′) and the unit rise is 5″ (unit length = 13″), 13 × 1.667 = 21.667 = 21¹¹⁄₁₆″.

Common difference (CD) in length of valley jacks is found the same way as with hip jacks: CD = (OC spacing ÷ 12) × unit length. See Figure 6-62 or look at the 3rd and 4th lines of a rafter square (see Figure 6-72).

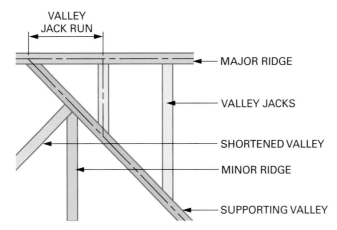

Figure 6-64 Determining the valley jack rafter run

Layout of the Valley Jack Rafter

The tip of the valley jack rafter where it abuts the ridge is laid out the same way as a common rafter. The tail is laid out the same way as a valley cripple jack (see Figure 6-66).

Calculation of the Valley Cripple Jack Rafters

Valley cripple jack rafters are special common rafters found in the same plane as the common rafters. These rafters connect on one end to a supporting valley rafter and on the other end to a shortened valley rafter (Figure 6-41). Their cheek cuts (tip and tail) are on the same side of the rafter. Figure 6-65 shows where the run can be measured. The run is *twice* the distance from the intersecting center of the valley rafters to the center of the valley cripple jack rafter. Use the formula **line length = unit length $\times$ 2 $\times$ run**. For example, calculate the line length of a valley cripple jack rafter if the run is 32″ (2.667′) and the slope is 12 on 12 (unit length = 16.97):

$$16.97 \times 2 \times 2.667' = 90.52'' = 90\tfrac{1}{2}''$$

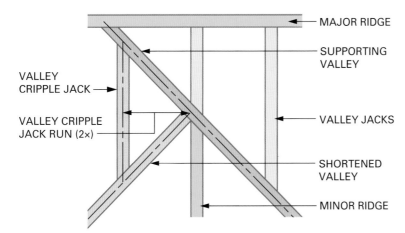

Figure 6-65 Measuring the valley cripple jack run

Layout of the Valley Cripple Jack Rafter

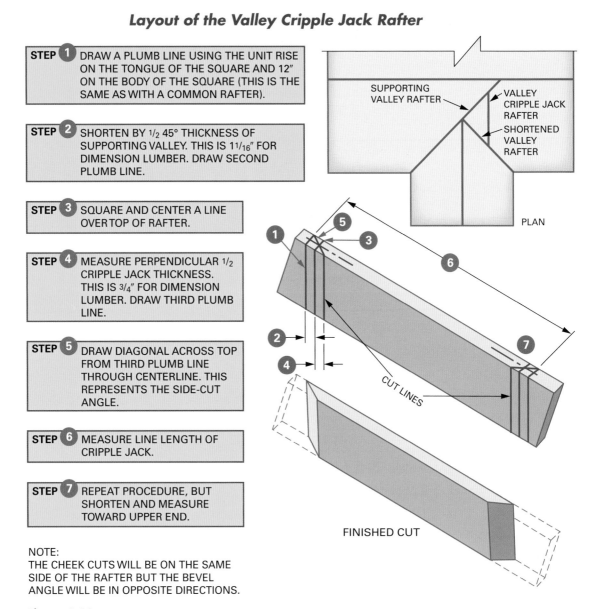

STEP 1 DRAW A PLUMB LINE USING THE UNIT RISE ON THE TONGUE OF THE SQUARE AND 12″ ON THE BODY OF THE SQUARE (THIS IS THE SAME AS WITH A COMMON RAFTER).

STEP 2 SHORTEN BY 1/2 45° THICKNESS OF SUPPORTING VALLEY. THIS IS 11/16″ FOR DIMENSION LUMBER. DRAW SECOND PLUMB LINE.

STEP 3 SQUARE AND CENTER A LINE OVER TOP OF RAFTER.

STEP 4 MEASURE PERPENDICULAR 1/2 CRIPPLE JACK THICKNESS. THIS IS 3/4″ FOR DIMENSION LUMBER. DRAW THIRD PLUMB LINE.

STEP 5 DRAW DIAGONAL ACROSS TOP FROM THIRD PLUMB LINE THROUGH CENTERLINE. THIS REPRESENTS THE SIDE-CUT ANGLE.

STEP 6 MEASURE LINE LENGTH OF CRIPPLE JACK.

STEP 7 REPEAT PROCEDURE, BUT SHORTEN AND MEASURE TOWARD UPPER END.

NOTE:
THE CHEEK CUTS WILL BE ON THE SAME SIDE OF THE RAFTER BUT THE BEVEL ANGLE WILL BE IN OPPOSITE DIRECTIONS.

SUPPORTING VALLEY RAFTER

VALLEY CRIPPLE JACK RAFTER

SHORTENED VALLEY RAFTER

PLAN

CUT LINES

FINISHED CUT

Figure 6-66 Laying out a valley cripple jack rafter

Calculation of the Hip/Valley Cripple Jack

Hip/valley cripple jack rafters are special common rafters found in the same plane as the common rafters. These rafters connect on one end to a hip rafter and on the other end to a valley rafter (see Figure 6-41). Their cheek cuts will be on *opposite* sides of the rafter. Figure 6-67 shows how to measure the run. Use the formula **line length = unit length × run.**

For example, calculate the line length of a hip/valley cripple jack if the run measures 65″ (5.417′) and the roof slope is 9/12 (unit length = 15):

$$15 \times 5.417 = 81.255 = 81\tfrac{1}{4}''$$

For layout procedures, see Figure 6-68.

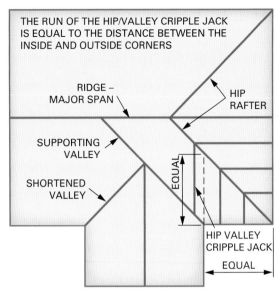

Figure 6-67 Determining the run of the hip/valley cripple jack rafter

Layout of the Hip/Valley Cripple Jack

STEP 1 DRAW A PLUMB LINE USING THE UNIT RISE ON THE TONGUE OF THE SQUARE AND 12″ ON THE BODY OF THE SQUARE; THIS IS THE SAME AS WITH A COMMON RAFTER.

STEP 2 SHORTEN ½ THE DIAGONAL 45° THICKNESS OF THE HIP RAFTER (1¹¹/₁₆″ IF USING 2× DIMENSION LUMBER) AND DRAW A SECOND PLUMB LINE.

STEP 3 SQUARE A LINE OVER THE TOP OF THE RAFTER FROM THE SECOND PLUMB LINE AND MARK THE CENTER.

STEP 4 MEASURE SQUARE TO THE PLUMB LINE ½ THE JACK THICKNESS AND DRAW A THIRD PLUMB LINE.

STEP 5 DRAW A DIAGONAL ACROSS THE TOP FROM THE TOP OF THE THIRD PLUMB LINE THROUGH THE CENTERLINE DRAWN IN STEP 3. **NOTE:** THIS CAN ALSO BE DRAWN USING THE SIDE-CUT NUMBERS FOUND ON A FRAMING SQUARE (FIGURE 6-50).

STEP 6 MEASURE AND MARK THE LINE LENGTH FROM THE TOP OF THE FIRST PLUMB LINE.

STEP 7 REPEAT THE PROCEDURE ON THE OPPOSITE END OF THE RAFTER BUT SHORTEN TOWARD THE TIP ½ THE DIAGONAL 45° THICKNESS OF THE VALLEY RAFTER AND HALF OF THE THICKNESS OF THE JACK.

NOTE:
CHEEK CUTS SHOULD BE ON OPPOSITE SIDES OF THE RAFTER; SEE THE FINISHED CUT. BE CAREFUL TO ORIENT THE DIRECTION OF THE CUTS CORRECTLY.

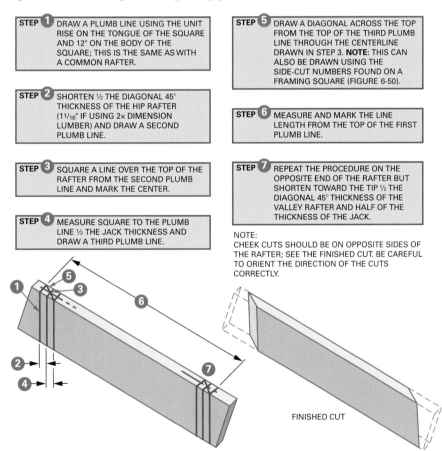

FINISHED CUT

Figure 6-68 Laying out a hip/valley cripple jack rafter

ASSEMBLY OF THE HIP ROOF

Ridgeboard Layout

> *The ridge length will not change if the slope of the roof is changed.*

Assembling the roof structure will be much easier if the ridgeboard has previously been laid out. See Figure 6-69.

STEP 1 LAY OUT THE DESIRED OC SPACING ALONG THE WALL PLATE (PROCESS NOT SHOWN).

STEP 2 MEASURE FROM THE CORNER ALONG THE PLATE ON THE LONG WALL ½ THE WIDTH OF THE BUILDING (THE COMMON RAFTER RUN); PLACE A MARK ON THE WALL PLATE.

STEP 3 SUBTRACT (BACK TOWARD THE CORNER) THE AMOUNT ADDED TO THE RIDGEBOARD AND PLACE A MARK ON THE PLATE. THIS MARK WILL CORRESPOND TO THE END OF THE RIDGEBOARD. SEE FIGURES 6-46 AND 6-47 FOR HIP RIDGE DETAILS.

STEP 4 REPEAT THIS PROCESS ON THE OTHER END OF THE BUILDING (PROCESS NOT SHOWN). THIS STEP IS NOT ABSOLUTELY NECESSARY BUT IT IS A GOOD METHOD TO DOUBLE-CHECK THE CALCULATED RIDGEBOARD LENGTH.

STEP 5 MEASURE FROM THE MARK IN STEP 3 TO THE NEXT OC COMMON RAFTER AND TRANSFER TO THE RIDGEBOARD (THE MARK MADE IN STEP 3 INDICATES ONE END OF THE RIDGEBOARD).

STEP 6 MARK REMAINING OC LAYOUT MARKS ON THE RIDGEBOARD (TRANSFER MARKS TO BOTH SIDES).

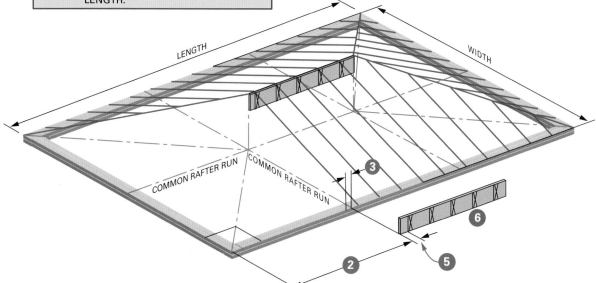

Figure 6-69 Laying out the hip ridgeboard

Assembly

The following is a suggested method for assembling a hip roof. This method recommends some common rafters being erected before the hip rafters. However, in the case of a square building (such as a garage), it may be easier to place the four hip rafters first. See Figure 6-70.

STEP 1 ERECT THE RIDGE WITH ONLY AS MANY COMMON RAFTERS AS NEEDED TO SUPPORT IT. A DIAGONAL BRACE WILL ALSO BE NECESSARY.

STEP 2 INSTALL THE HIP RAFTERS. THE HIP RAFTERS EFFECTIVELY BRACE THE ASSEMBLY.

NOTE: THIS PROCEDURE SHOWS THE HIPS FRAMED TO THE RIDGE. IF HIPS ARE FRAMED TO THE COMMON RAFTERS, INSTALL THE COMMONS TO THE END OF THE RIDGE BEFORE THE HIPS ARE PLACED.

STEP 3 INSTALL THE REMAINING COMMON RAFTERS IN PAIRS OPPOSING EACH OTHER. SIGHT THE TOP EDGE OF THE RIDGE FOR STRAIGHTNESS AS FRAMING PROGRESSES.

STEP 4 INSTALL THE HIP JACK RAFTERS IN PAIRS OPPOSING EACH OTHER. SIGHT THE HIP FOR STRAIGHTNESS AS JACKS ARE INSTALLED.

NOTE: IT IS BEST TO FIRST INSTALL A PAIR OF JACKS ABOUT HALFWAY UP THE HIP TO STRAIGHTEN IT.

ROOF FRAMING PLAN

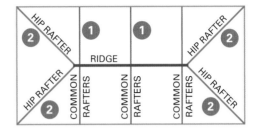

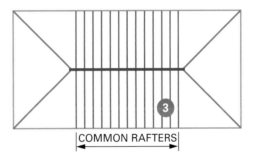

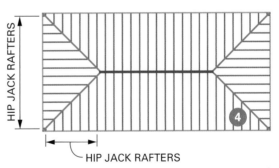

Figure 6-70 Erecting the hip roof frame

Sheathing the Hip Roof

Assume $4' \times 8'$ sheathing is used.

- **Step 1**—Measure up from the fascia 48¼″ and strike a chalk line along the tops of the rafters. **Note:** Roof sheathing should always be placed to a line, not simply flush with the fascia; a slight wave in the fascia can cause misalignment and make it difficult to evenly place subsequent sheets (Figure 6-71).

- **Step 2**—Establish a square line on the top (center) of one of the rafters (8′ or less from the corner).

TIP

By striking the chalk line 48¼″ up instead of 48″ up will allow for a slight variation in the fascia.

TIP

It is generally helpful to place layout marks on the top of the sheet (uphill edge). Bowed rafters can be moved slightly to align with the intended layout.

- **Step 3**—Use the line drawn in Step 2 and measure at each end the long point (bottom) and short point (top) dimensions of the cut. **Note:** Some carpenters let a portion of the sheathing overhang the hip rafter, and cut it later. However, sheathing ending in a valley must accurately be cut before placement.

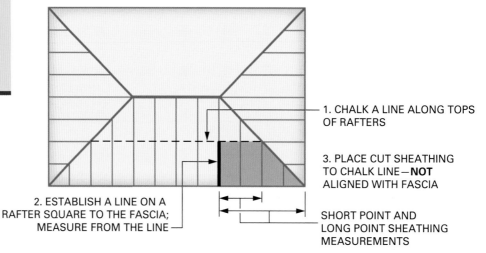

1. CHALK A LINE ALONG TOPS OF RAFTERS

3. PLACE CUT SHEATHING TO CHALK LINE—**NOT** ALIGNED WITH FASCIA

2. ESTABLISH A LINE ON A RAFTER SQUARE TO THE FASCIA; MEASURE FROM THE LINE

SHORT POINT AND LONG POINT SHEATHING MEASUREMENTS

Figure 6-71 Sheathing a hip roof

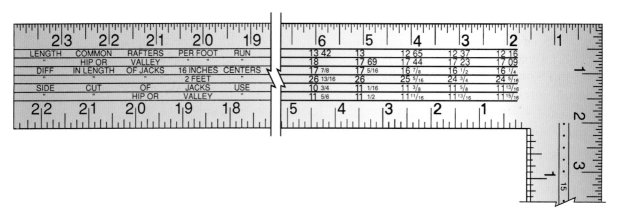

Figure 6-72 The rafter square

THE RAFTER SQUARE

As mentioned previously, the rafter square with embossed rafter tables has many uses and contains helpful information for the carpenter.

The first line shows "Length of Common Rafters per Foot of Run" (also known the unit length). Under each of the measured numbers is a value. For example, under the 4″ mark is the number 12.65, which refers to the unit length of a 4/12 roof slope.

The second line (17.44) shows "Hip or Valley Rafters per Foot of Run" (hip/valley unit lengths) (see Table 6-1).

The third line (16⅞″) shows "Difference in Length of Jacks 16″ OC″ (CD 16OC).

The fourth line (25⁵⁄₁₆″) shows "Difference in Length of Jacks 2 Feet OC" (CD 24 OC) (see Figure 6-62).

The fifth line (11⅜″) shows "Side Cut of Jacks Use." See Figure 6-50.

The sixth line (11¹¹⁄₁₆″) shows "Side Cut of Hip or Valley Use." See Figure 6-50.

COMPLEX ROOF HIP/VALLEY AND RIDGE INTERSECTIONS

When a house has a complex hip/valley ridge system, there are specific procedures to follow to accurately determine lengths of the various ridges, and to aid in precise placement of other components. Figure 6-73 A–E illustrates details that are commonly encountered in complex roof framing.

Calculating Ridge Lengths

The ridge length does not depend on the slope of the roof. To calculate the lengths of ridges R1, R2, R3, and minor ridge M1 (Figure 6-73), first the lengths are found to the centerline intersections (theoretical length) and then the necessary adjustments are added or subtracted to determine the actual length.

R1 = length of the building minus width of the building, plus adjustments.

$$\textbf{R1} = (54' - 28') + ¾'' + ¾'' = \textbf{26' 1½''}$$

Notice a common rafter abuts each end of the ridge. Use the diagram in Figure 6-73A.

R2 = half of the 12′ span + 4′ (the amount the 12′ section of building projects out from the main building) minus adjustments. Note the pink triangle to the right of R2; this illustrates the two legs of the triangle are the same length (6′); this is always the case for a hip or valley at a 45° angle to the building. Detail Figure 6-73C shows that 1¹⁄₁₆″ needs to be subtracted from the theoretical ridge length:

$$\textbf{R2} = 6' + 4' - 1¹⁄₁₆'' = \textbf{9' 10¹⁵⁄₁₆''}$$

R3 = 6′ plus and minus adjustments. R3 connects to the main ridge at one end (with valley rafters also) and has hip rafters connecting with it at the other. Notice that the hip and valley are parallel to each other. The theoretical ridge length, the distance between the hip and valley at the ridge, which is 6′, can be found at the lower-left edge of the building. Detail Figure 6-73B will be used for the adjustment at the hip, and detail Figure 6-73D will be used for the adjustment at the ridge. Thus, the actual length is:

$$\textbf{R3} = 6' + 1¹³⁄₁₆'' - ¾'' = \textbf{6' 1⁵⁄₁₆''}$$

M1 is a minor ridge or a portion of R1. The dimension sought is to mark R1 where the supporting valley rafter's edge (long point) contacts the ridge. This will help in locating the valley rafter during construction. The theoretical dimension of M1 is found by noting the distance between the supporting valley rafter and the hip rafter at the upper-right edge of the building (10′). The adjustment made at the ridge/valley intersection (+⁵⁄₁₆″) is illustrated in Figure 6-73E, and at the other end of M1, by Figure 6-73A.

$$\textbf{M1} = 10' + ⁵⁄₁₆'' + ¾'' = \textbf{10' 11⁄₁₆''}$$

All of this information can be mathematically derived; however, knowing how to use the square can prevent extra calculations.

In the case where lumber of different thicknesses is being used simultaneously, or if all lumber is of a different thickness (than 1½″), then the diagrams (Figure 6-73) will need to be modified slightly. Keep this in mind especially if combining engineered lumber with dimension lumber.

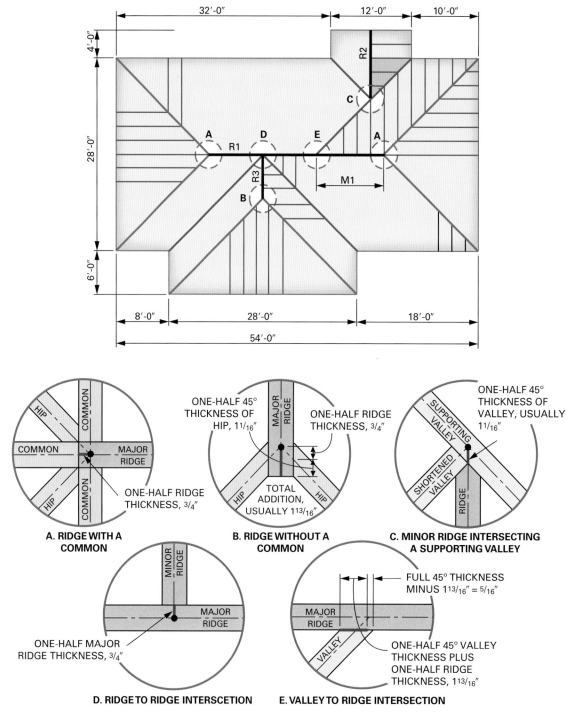

Figure 6-73 Ridgeboard lengths are adjusted depending on the style of the framing

Broken Hip

The *broken hip* is a framing member that can add versatility to hip roof construction. Figure 6-74 illustrates an "L"-shaped building where each leg of the "L" is the same width; this allows the hips (and valley) to meet and a broken hip is not necessary.

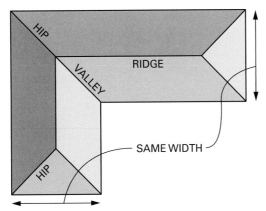

Figure 6-74 Hip ridges meet if ends of the building are the same width

It isn't always possible to build a structure with intersecting roof sections of the same width. Figure 6-75 shows one section of the building wider than the other. The ways to accommodate this are:

1. Frame the roof sections at different slopes while keeping the ridges at the same elevation (generally a poor idea).

2. Frame the sections to slope the same but with a small gable end sticking up above the smaller (narrower) roof section (see Figure 6-76).

3. Frame all roof sections to slope the same *and* use a broken hip to connect them. Figure 6-75 shows a top view, two elevation views, and a framing detail illustrating the broken hip.

Notice that the broken hip also runs at a 45° angle to the outside wall. Also notice (see the **detail** of Figure 6-75) the small triangle formed by the inside edge of the broken hip, the upper portion of the hip, and the ridge; the length of the broken hip is identical to the portion of the hip rafter measuring from one ridge (lower) to the next (upper).

If all roof sections are framed to the same slope and not connected with a broken hip, then a small triangular gable end will show and need to be sided (Figure 6-76).

The broken hip rafter allows houses with very complex shapes to be framed with hip roofs. This subject is challenging enough, that going beyond the basics covered in this book takes a great deal of supplemental information and practice. If one would like to explore this subject further, a book dedicated specifically to roof framing is recommended.

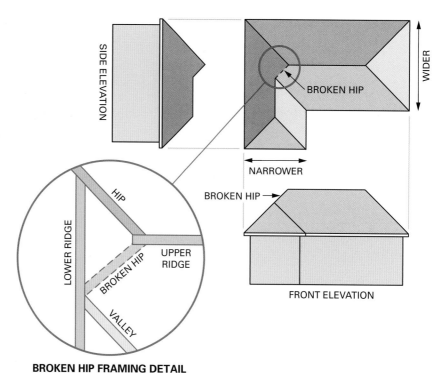

Figure 6-75 A broken hip is necessary to make ridges of different elevations meet

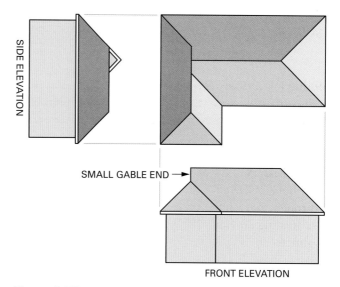

Figure 6-76 Without a broken hip, a small gable end is necessary

Cathedral Ceiling

Many homes have high, open, sloped cathedral ceilings without collar ties. To frame such a ceiling, special details must be followed to ensure the structural integrity of the building. If no precautions are taken, the downward forces (weight) will cause the ridge to sag and that, in turn, may force the walls to bow outward (see Figure 6-77).

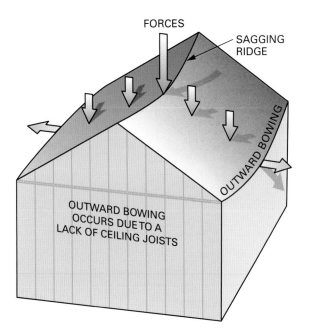

Figure 6-77 Precautions must be taken when building a cathedral ceiling

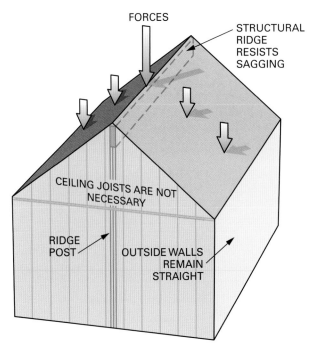

Figure 6-78 The ridge must be strong enough to prevent deflection

By utilizing an adequate ridge that is supported at both ends, the rafters in effect "hang" from the ridge instead of pulling it downward. By preventing the ridge from sagging, the walls will not be forced outward. See Figure 6-78.

> *Sizing of a key structural member should always be done by a qualified professional.*

Ventilation

This brief introduction to roof ventilation will help the reader to understand the placement options of rafters against the ridge. In order for buildings to "breath" properly and avoid moisture build-up, it is necessary to provide adequate ventilation. Roofs repeatedly get very hot and then cool off, causing humid air, or even some condensation to form. To keep the roof cooler and channel the moist air away, air flow is necessary. Typically, roof ventilation starts at the soffit. As the roof heats up, the hot air rises and exits at the ridge via a ridge vent, through gable end vents, or through roof vents. As the air exits, it draws cooler air up through the soffit vents, creating air flow. This process removes moisture and keeps the roof cooler (Figure 6-79).

Rafter Placement at the Ridge

There are two common methods for placement of the rafters against the ridge.

- **Method 1**—This method works with a single 2× ridge beam, or if a ridge vent is not utilized. The roof sheathing is cut back from the ridge to create a gap (1½″ or so) for air to escape. A ridge vent is wide enough to cover the gap and weatherproof the roof (Figure 6-80).
- **Method 2**—This method is often necessary if ridge ventilation is needed while framing with a large structural ridge (Figure 6-81).

> *Rafters should never be allowed to extend unsupported below the face of the ridge. Codes also generally require that the ridge depth not be less than the end of the cut of the rafter.*

The examples shown here are by no means all options for framing at the ridge. Some prefer to place the rafters on top of a structural ridge. This way, the rafters extend over the ridge and the venting is easy to accomplish.

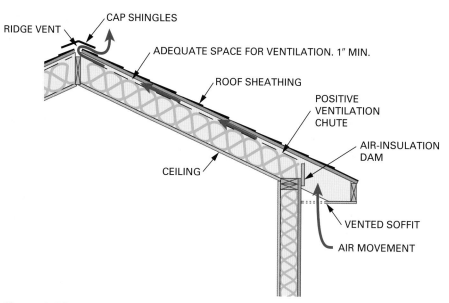

RIDGE VENT
CAP SHINGLES
ADEQUATE SPACE FOR VENTILATION. 1″ MIN.
ROOF SHEATHING
POSITIVE VENTILATION CHUTE
AIR-INSULATION DAM
CEILING
VENTED SOFFIT
AIR MOVEMENT

Figure 6-79 A common method of providing ventilation

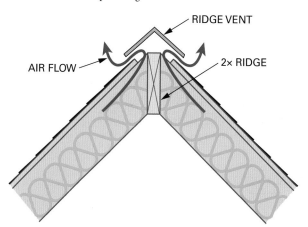

RIDGE VENT
AIR FLOW
2× RIDGE

Figure 6-80 Framing the top of the rafter flush with the top of the ridge

If the sheathing was not cantilevered over the rafters and it was cut back as in Method 1, then the ridge vent would probably not be wide enough to cover the ridge and vent space.

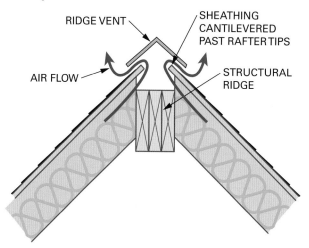

RIDGE VENT
SHEATHING CANTILEVERED PAST RAFTER TIPS
AIR FLOW
STRUCTURAL RIDGE

Figure 6-81 Framing the top of the rafter above the top of the ridge

Notice in Figure 6-81 the tops of the rafters extend above the ridge approximately 1½″. The sheathing is then cantilevered over the rafter tips *without* blocking the air flow from between the rafters.

Calculating Ridge-Post Length

This method is necessary when constructing a cathedral ceiling requiring a heavy structural ridge beam. By accurately calculating the ridge-post height, the ridge can be placed *before* any rafters (see Figure 6-82). Then rafters can be placed one at a time.

- **Step 1**—Find the adjusted run. To find this, subtract half of the width of the ridge from the total run. This allows one to calculate line length and total rise to a point on the *face* of the ridge (instead of the middle of the ridge).
- **Step 2**—Calculate adjusted line length.

    Unit length × adjusted run = adjusted rafter (line) length.

    Adjusted rafter length is measured from the ridge plumb cut to bird's mouth plumb line.

- **Step 3**—Calculate adjusted total rise.

    Adjusted total rise = adjusted run × unit rise

    This is measured from the top of the top plate at the back of the bird's mouth cut (seat plumb line), vertically, to a similar part on the rafter at the ridge plumb cut. The dashed diagonal line on the rafter shows the relative location of this measurement in Figure 6-82.

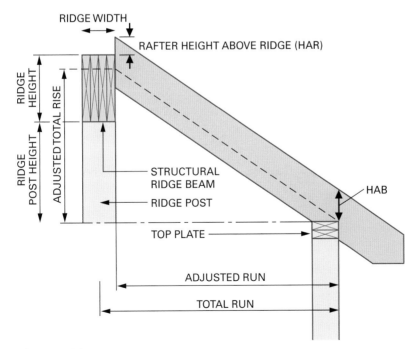

Figure 6-82 Calculating ridge-post length

> *Remember, with a structural ridge beam, the rafter tip extends above the ridge in order to provide a ventilation channel.*

- **Step 4**—Add the height above the bird's mouth (HAB) to the adjusted total rise. This will elevate the total rise to the tip of the rafter (a known point).
- **Step 5**—Add the ridge height and the height above ridge (HAR) together. This yields the measurement from the tip of the rafter to the bottom of the ridge.
- **Step 6**—Subtract the sum found in Step 5 from the sum found in Step 4; this will give the ridge-post length from the top of the top plate upward. Eventually, there should be a support post extending from the structural ridge to the foundation.

Example Problem: Using the given information, calculate the ridge-post length.
Given: Ridge width = 6″, ridge height = 14″, total span = 32′, unit rise = 5″, HAB = 7″, HAR = 1½″. This dimension can vary depending on ridge vent type and roof size.

- **Step 1**—Adjusted run = ½ building span (16′) minus ½ ridge width (3″); therefore:

$$16' - 3'' = 15' \ 9'' \text{ or } \mathbf{15.75'} = \textbf{adjusted run}$$

- **Step 2**—Unit length × adjusted run = adjusted line length; therefore:

$$13 \times 15.75' = \mathbf{204.75''} = \textbf{adjusted line length}$$

- **Step 3**—Adjusted total rise = total run (adjusted) × unit rise; therefore:

$$\text{adjusted total rise} = 15.75 \times 5'' = \mathbf{78.75''} = \textbf{total rise}$$

This is measured to a place parallel to the face of the ridge—not at the center of the ridge (half of the width of the ridge was subtracted in Step 1).

- **Step 4**—Add the HAB to the adjusted total rise,

$$7'' + 78.75'' = \mathbf{85.75''}$$

- **Step 5**—Add ridge height + HAR,

$$14'' + 1\frac{1}{2}'' = \mathbf{15\frac{1}{2}''}$$

- **Step 6**—Find ridge-post length,

$$85.75'' - 15.5'' = \mathbf{70.25''} = \textbf{ridge-post length}$$

This calculated ridge-post length is sized to be placed on the top plate. The post can extend from the floor/foundation (recommended) to the bottom of the ridge by adding the height of the wall to the ridge-post length. The ridge must be supported at both ends.

Fastening Hardware

In some locations, high winds, earthquakes, and other problems require the use of specialized hardware for roof construction. Figure 6-83 and Figure 6-84 show some of this hardware. Always check local code requirements.

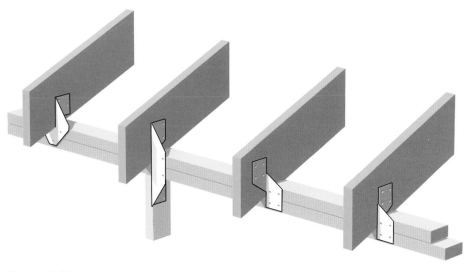

Figure 6-83 Special hardware for roof construction

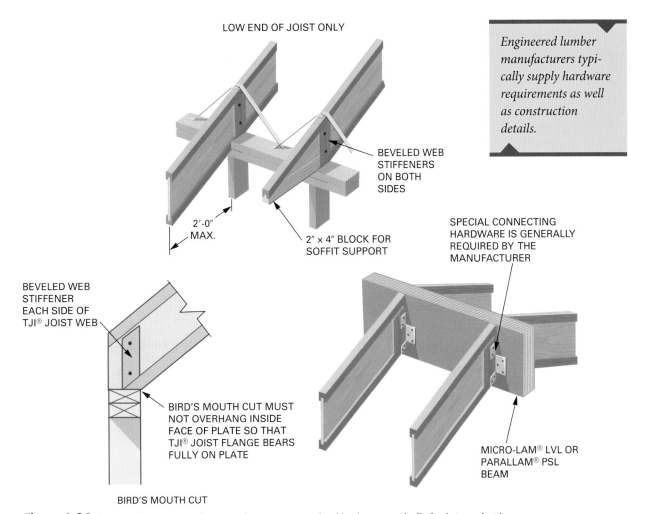

LOW END OF JOIST ONLY

BEVELED WEB
STIFFENERS
ON BOTH
SIDES

Engineered lumber manufacturers typically supply hardware requirements as well as construction details.

2'-0"
MAX.

2" × 4" BLOCK FOR
SOFFIT SUPPORT

SPECIAL CONNECTING
HARDWARE IS GENERALLY
REQUIRED BY THE
MANUFACTURER

BEVELED WEB
STIFFENER
EACH SIDE OF
TJI® JOIST WEB

BIRD'S MOUTH CUT MUST
NOT OVERHANG INSIDE
FACE OF PLATE SO THAT
TJI® JOIST FLANGE BEARS
FULLY ON PLATE

MICRO-LAM® LVL OR
PARALLAM® PSL
BEAM

BIRD'S MOUTH CUT

Figure 6-84 Engineered lumber manufacturers often require specialized hardware specifically for their product line

UNEQUAL SLOPING ROOFS

It is necessary at times to frame two differently sloping roofs opposite each other. Houses built this way generally have the soffits built to the same height. This helps maintain a uniform look. However, due to the different roof slope angles, projections of the roofs are different.

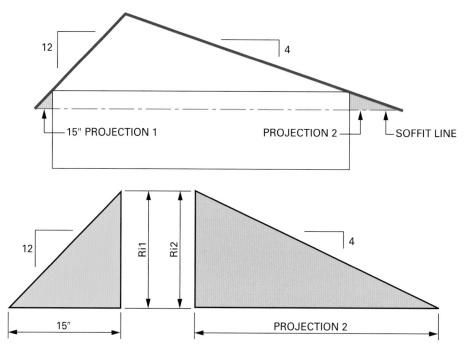

Figure 6-85 Unequally sloping roofs

If the slopes of the roofs are known, or can be measured, and the projection on one side is known, then it is relatively easy to determine the projection needed on the other side.

Figure 6-85 shows a 12 on 12 slope on the left side and a 4 on 12 slope on the right side. The projection on the left side is 15″ (1.25′) and the projection on the right is unknown.

The total rises of each of the rafters are equal and the same can be said for the rise of the overhangs (Ri1 = Ri2). Using the formula **unit rise × run** (in feet) = **total rise**, Projection 2 can be calculated. In the formula, substitute the projection dimension for run.

First calculate Ri1:

$$\text{Unit rise} \times \text{projection} = \text{total rise}$$
$$12 \times 1.25 = 15″$$

If Ri1 = Ri2, then Projection 2 can be calculated using a variation of the same formula:

$$\text{Projection 2} = \text{total rise} \div \text{unit rise.}$$
$$\text{Projection 2} = 15 \div 4 = \textbf{3.75′} = \textbf{45″}$$

Measuring Roof Slope

During a remodeling project where one may need to match an existing roof slope, it is important to be able to measure the slope of the roof (Figure 6-86).

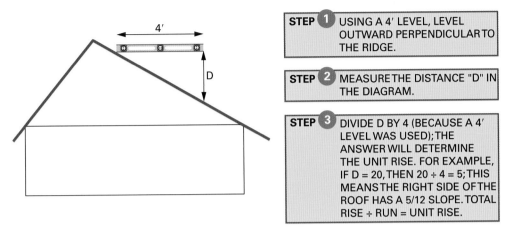

STEP **1** USING A 4' LEVEL, LEVEL OUTWARD PERPENDICULAR TO THE RIDGE.

STEP **2** MEASURE THE DISTANCE "D" IN THE DIAGRAM.

STEP **3** DIVIDE D BY 4 (BECAUSE A 4' LEVEL WAS USED); THE ANSWER WILL DETERMINE THE UNIT RISE. FOR EXAMPLE, IF D = 20, THEN 20 ÷ 4 = 5; THIS MEANS THE RIGHT SIDE OF THE ROOF HAS A 5/12 SLOPE. TOTAL RISE ÷ RUN = UNIT RISE.

Figure 6-86 Determining the slope of an existing roof

DORMERS

Dormers are especially popular on steeper-sloping roofs where natural lighting and/or upper-floor living space is desired. It is common to add dormers to an existing structure; however, it is easier to plan them into the original design. The two most popular types are shed dormers and gable dormers.

Gable Dormers

Gable dormers (Figure 6-87) have a small gable roof with a window located under the gable. The top of the window is generally placed the same height above the floor as other windows in the structure. There may be exceptions due to decorative windows being used, such as a round-top window, or there may be ceiling height limitations. Gable dormers can be framed using different methods; following is one suggested method.

- **Step 1—Determine the desired interior width of the dormer and frame the roof opening accordingly.** The rafters must be reinforced on each side to carry the added load. It will be necessary to double or even triple these rafters or possibly use engineered lumber.

- **Step 2—Determine the length of the opening, place the headers, and then frame in the shortened rafters above and below the headers.** Often there is a knee wall located under the sloped roof, blocking off an area that is generally unusable (see Figure 6-87a). This knee wall will occasionally determine the location of the front (gable) wall of the dormer. However, the wall can be

Framing dormers on a steep roof can be dangerous—make sure to use appropriate staging and fall protection.

It is assumed that the reader at this point understands the basics of roof framing; therefore, explanations may be limited. If additional instruction is necessary such as layout of rafters, refer back to the appropriate section.

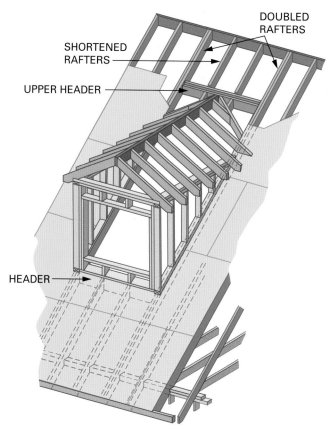

Figure 6-87 Gable dormer

placed wherever desired, but it is often placed directly over the outside wall of the structure. The upper header location is determined by the desired headroom, the type of framing used, and the slope of the roof.

If the dormer will have a flat interior ceiling, the bottom of the upper header can be placed even with the top of the sidewalls of the dormer and the ceiling joists can rest on the sidewalls of the dormer. If a cathedral ceiling is desired, the upper header is placed with the bottom flush with the bottom of the dormer's ridge, as in Figure 6-87.

- **Step 3—Build the sidewalls.** The sidewall height will have been determined in the previous steps. The walls are built directly over the reinforced rafters at the edges of the opening. Cut the sidewall bottom plates (one for each side); leave room for the doubled top plates to be added. Each end of the bottom plate will have to be beveled to match the roof slope (see Figure 6-87a). Fasten the sidewall bottom plates to the roof flush with the opening.

  Bevel one end of the top plate to match the roof slope and place the beveled end on top of the upper end of the sidewall plate. Level outward and

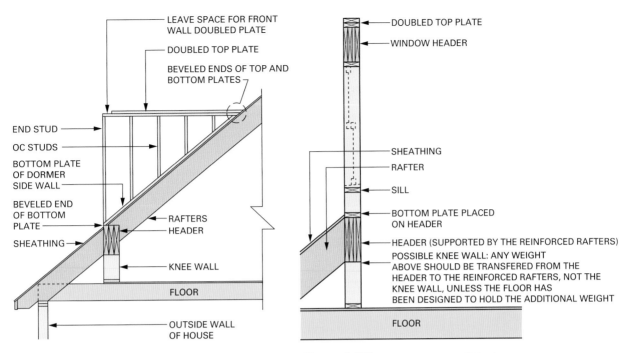

Figure 6-87a Side view of a gable dormer (no roof)

Figure 6-87b Side view of front wall detail

measure down to find the length of the end stud. At the same time, mark the length of the top plate (extra hands will help with this). Cut the top plate and the end stud and nail in place; brace temporarily. Lay out the wall plates; cut and place the remaining sidewall studs. Add a doubled plate, making sure to leave room at the end for the front (gable) wall's doubled plate to overlap. Make sure both sidewalls have identical measurements.

- **Step 4—Build the front wall.** Cut a bottom and top plate to fit between the two outside walls. Lay out the window location and the stud positions. The bottom plate will be placed perpendicular to the roof slope (see Figure 6-87b). Build the front wall.

  Securely fasten the walls to each other and to the roof structure. Add a doubled plate making sure to overlap the two end walls.

- **Step 5—Cut the common rafters and place the ridge.** One end of the ridge can attach to the header (calculated and placed earlier) and the other will overhang the gable end of the dormer. Using half of the width of the dormer as the run, calculate, lay out, and cut as many common rafters as are necessary for the dormer (also cut the two fly rafters if there is a gable projection). Erect the ridge and rafters. On each side of the dormer where the ridge meets the header, cut small valley rafters. Using half of the width of the dormer as the run, calculate and cut the small valley rafters. **Note:** There will be double cheek cuts on the tips (see Figure 6-55) because the valleys are meeting at

TIP

Some of the steps above such as lengths of plates and studs can be derived mathematically; however, it is advisable to build a dormer or two to gain experience before attempting to build a dormer based on theoretical calculations.

the intersection of two framing members (a ridge and a header); there is no bird's mouth so the tail will be shortened similar to that of a valley cripple jack (see Figure 6-66). The rest of the valley jack rafters can then be cut and placed (see Figure 6-64).

There are times when the upper portion of the dormer is framed directly on existing roof sheathing. If this technique is desired, Figure 6-88 illustrates how to cut this special type of valley jack rafter.

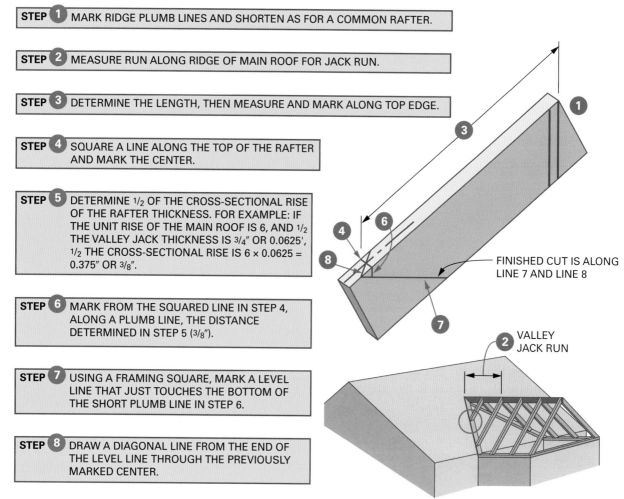

STEP 1 MARK RIDGE PLUMB LINES AND SHORTEN AS FOR A COMMON RAFTER.

STEP 2 MEASURE RUN ALONG RIDGE OF MAIN ROOF FOR JACK RUN.

STEP 3 DETERMINE THE LENGTH, THEN MEASURE AND MARK ALONG TOP EDGE.

STEP 4 SQUARE A LINE ALONG THE TOP OF THE RAFTER AND MARK THE CENTER.

STEP 5 DETERMINE 1/2 OF THE CROSS-SECTIONAL RISE OF THE RAFTER THICKNESS. FOR EXAMPLE: IF THE UNIT RISE OF THE MAIN ROOF IS 6, AND 1/2 THE VALLEY JACK THICKNESS IS 3/4" OR 0.0625', 1/2 THE CROSS-SECTIONAL RISE IS 6 × 0.0625 = 0.375" OR 3/8".

STEP 6 MARK FROM THE SQUARED LINE IN STEP 4, ALONG A PLUMB LINE, THE DISTANCE DETERMINED IN STEP 5 (3/8").

STEP 7 USING A FRAMING SQUARE, MARK A LEVEL LINE THAT JUST TOUCHES THE BOTTOM OF THE SHORT PLUMB LINE IN STEP 6.

STEP 8 DRAW A DIAGONAL LINE FROM THE END OF THE LEVEL LINE THROUGH THE PREVIOUSLY MARKED CENTER.

FINISHED CUT IS ALONG LINE 7 AND LINE 8

VALLEY JACK RUN

NOTE:
THE END OF THE VALLEY JACK MAY NOT EXTEND TO THE EDGE OF THE ROOF. HOWEVER, EVEN IF IT ENDS ON A DORMER WALL, THE CUT DETAIL (STEPS 4–8) WILL BE THE SAME.

Figure 6-88 Laying out a valley jack for a sheathed roof

Shed Dormers

This is different from the gable dormer primarily in external appearance. Many of the techniques used to frame the shed dormers (Figure 6-89) have been previously covered in gable dormers (essentially Steps 1–4); therefore this section will add only supplemental information.

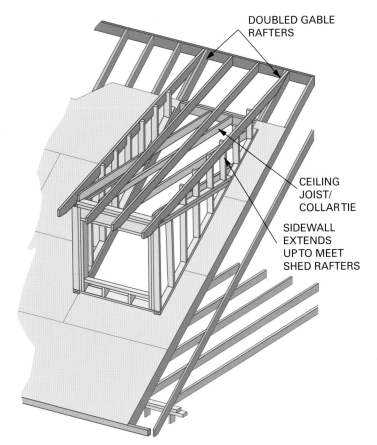

DOUBLED GABLE
RAFTERS

CEILING
JOIST/
COLLAR TIE

SIDEWALL
EXTENDS
UP TO MEET
SHED RAFTERS

Figure 6-89 Shed dormer

> There are structural
> implications to build-
> ing large shed dormers
> and careful engineer-
> ing must be adhered
> to. Large dormers
> can place a great deal
> more stress on the rest
> of the roof structure.

Occasionally, shed dormers are quite large and extend the total length of a structure. When the entire side of a structure is dormered, the front wall is placed directly over the outside wall of the main building. This type of a dormer will likely require a structural ridge.

Shed dormers are simpler to build mainly due to their roof structure being less intricate. Follow the layout process in Figure 6-20 for common rafters. The rafter run is measured from the front of outside wall of the dormer, horizontally, to the face of the ridge. The shed roof rafters will not match the slope of the main roof. The sidewalls of the shed dormer must be extended up to meet the bottom of the sloping shed rafters. Ceiling joists can either end at the upper header placed between the main roof rafters or, if long enough lumber can be acquired, ceiling joists can attach from the front wall of the dormer on one side of the structure to the rafters on the opposite side of the roof.

SKYLIGHTS

Framing openings for skylights is a simple process and is nearly identical to framing floor openings. Before starting the framing process, the skylight must be selected and the rough opening (RO) determined. The manufacturer will supply

TIP

> Some carpenters pre-
> fer to build the front
> wall first and then
> place the shed rafters,
> followed by the side-
> walls. This method
> allows the sidewall
> studs to extend in one
> piece from the main
> roof to the shed
> rafters.

the RO with the skylight. Depending on the size of the unit, extra rafters may be necessary on each side of the skylight to help carry weight from the headers and the skylight itself. See Figure 6-90.

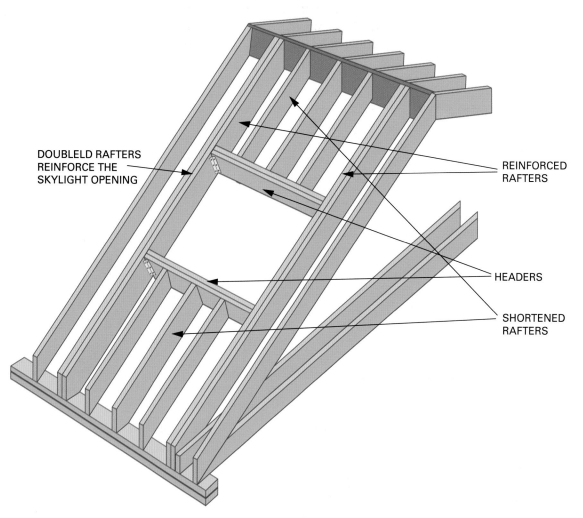

DOUBLELD RAFTERS REINFORCE THE SKYLIGHT OPENING

REINFORCED RAFTERS

HEADERS

SHORTENED RAFTERS

Figure 6-90 Framing typical of a skylight opening

Once the size and location of the skylight have been determined, reinforce the rafters on each side of the opening. Next place the header. Fill in the shortened rafters, and then sheath the roof, making sure to trim the sheathing flush with the opening.

This book will not discuss the installation of the skylights; each manufacturer has their preferred methods. Specific flashing kits and other hardware are sold with skylights. Installation information is supplied when purchasing a unit. Make sure to select the skylight before starting framing; some manufacturers have specific slope requirements and may require a special curb to help the unit to shed water.

There are some interior details to consider. If the unit is placed in a cathedral ceiling, there will be minimal obstructions from the surrounding structure. All it needs is to be trimmed. This is the simplest installation See Figure 6-91.

It is possible to splay all four sides of the shaft to maximize light distribution.

Insulating the entire shaft is very important, especially in cold climates.

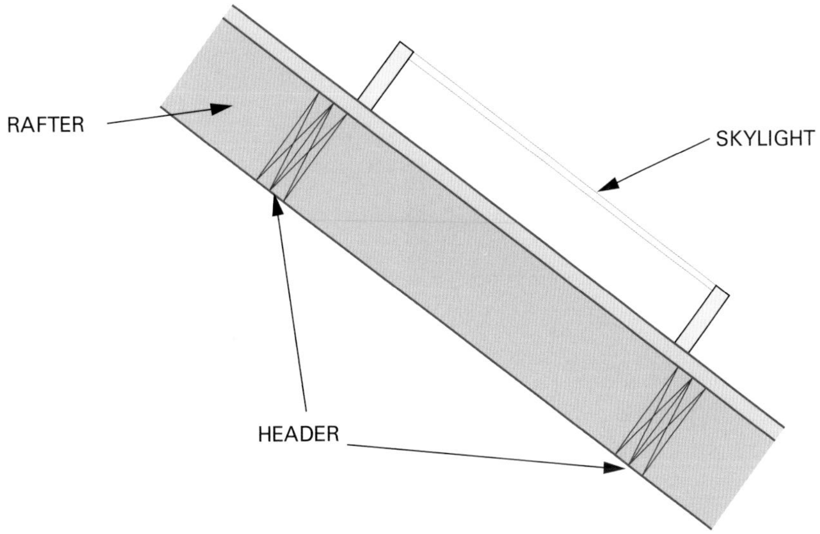

Figure 6-91 A skylight in a cathedral ceiling

Another installation variation (Figure 6-92) shows the skylight shaft penetrating an attic space. A rectangular shaft is built through the attic to the ceiling beneath. The shaft is square to the slope of the roof. This method does not allow as much light to reach the living space as the method in Figure 6-93.

A third variation (Figure 6-93) shows the skylight shaft penetrating an attic space; however, this method shows a splayed shaft, which allows a better spread of light into the living space. The uphill side is perpendicular to the roof slope while the lower side of the shaft is framed plumb. This includes framing the header plumb. The penetration through the ceiling joists will also need reinforcement such as doubled ceiling joists and headers.

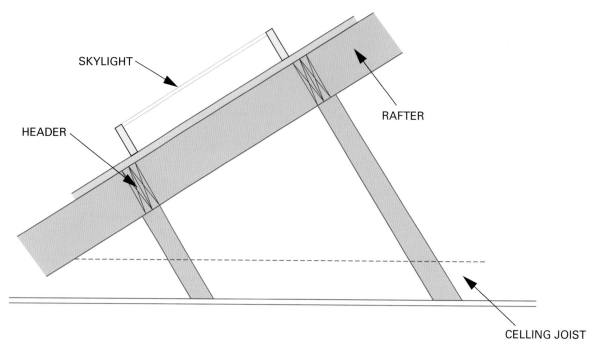

Figure 6-92 A skylight with a rectangular shaft

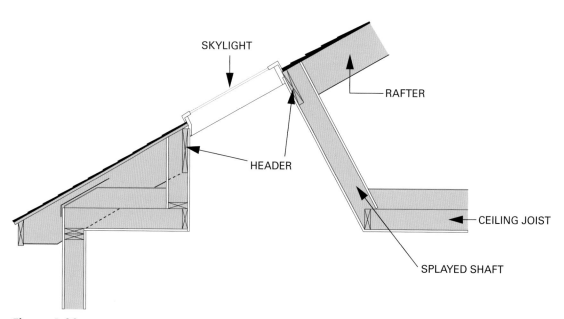

Figure 6-93 A skylight with a splayed shaft

OCTAGON ROOF FRAMING

Angles used on octagon-shaped roof structures are also found on bay window roofs, so they are fairly common, thus the need to understand their framing. When framing an octagon-shaped roof, typically hip rafters are placed first, followed by commons and jacks.

One method of framing requires many of the rafters to have steep side-cut angles (difficult!) where they join. Another method does not require side cuts at all. Still another, may be a combination of both methods. Regardless, all have the same details at the bird's mouth and at the tails.

Side Cut Method (More Difficult)

Figure 6-94 illustrates the hip rafters being placed in opposing pairs; each pair is identical. Note that three different-sized hip rafters are needed. Rafter #1s are full length, do not need side cuts, and do not need to be shortened. Rafter #2s do not need side cuts, but do need to be shortened. Rafter #3s (there are four #3s) need to be shortened *and* they have double side cuts. Rafter #4 represents a common rafter which needs to be shortened *twice* and needs double side-cut angles.

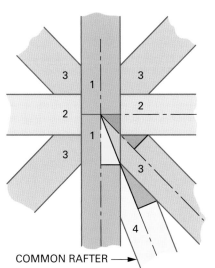

ASSUME ALL MATERIAL IS 2× (1½" WIDE)

RAFTERS MARKED "1" ARE PLACED FIRST; THEY ARE NOT SHORTENED.

#2s ARE PLACED SECOND AND NEED TO BE SHORTENED ½ THE WIDTH OF #1s (¾").

#3s ARE SHORTENED ½ THE DIAGONAL THICKNESS OF #1s (11/16" = HYPOTENUSE OF THE BLUE TRIANGLE), AND ½ OF THE THICKNESS OF HIP #3 (¾" = ONE LEG OF RED TRIANGLE).

#4, COMMON RAFTERS, ARE SHORTENED TWICE: ½ THE 22.5° THICKNESS OF #1s (1 5/16" HYPOTENUSE OF THE YELLOW TRIANGLE), AND THE LENGTH OF THE LONGER LEG OF SAME TRIANGLE (1 13/16" GREEN TRIANGLE).

COMMON RAFTER →

Figure 6-94 Framing an octagon roof using side cuts

Block Method (Easier)

This method allows one to build an octagon structure without *any* side cuts. Figure 6-95 illustrates the octagon hip rafters joining at an octagon-shaped block.

In this case, all eight hip rafters are the same size and need to be shortened ½ of the width of the center block. Instead of the common rafters touching the hip rafters as in Method 1, horizontal blocking is placed between the hip rafters (see Figure 6-96). The common rafters attach to the blocking. Another set of blocking is placed for the jack rafters. The jack rafters also do not need side-cut angles.

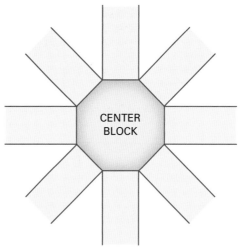

ALL HIP RAFTERS SIMILAR,
NO SIDE-CUT ANGLES

Figure 6-95 Framing an octagon roof with a center
block

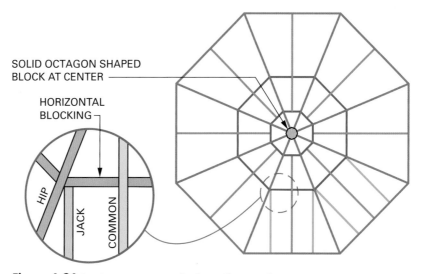

SOLID OCTAGON SHAPED
BLOCK AT CENTER

HORIZONTAL
BLOCKING

HIP

JACK

COMMON

Figure 6-96 Framing an octagon roof without side-cut angles

Octagon Calculations

In order to construct octagon roofs, it is essential to understand some of the necessary calculations. There are different methods of performing the calculations, but
here the Pythagorean Theorem ($a^2 + b^2 = c^2$) will be used because most carpenters
are familiar with it.

Octagon Site Layout

Example Problem: Lay out an octagon gazebo with each side measuring 6′. How
wide is the gazebo?

Break up the octagon into smaller pieces (Figure 6-97). Notice that the exterior
sides are the same length as the sides of the interior square.

Also notice that **a** = **b** and that they are two legs of a right triangle; thus if **c** is known, then **a** and **b** can easily be found. $a^2 + b^2 = c^2$, or in this case $2a^2 = c^2$. Follow the solution in Figure 6-98. The overall width, W, is found by adding 2(a) + 6. As Figure 6-98 shows there are two **a**'s along the width of the octagon plus the width of the center square:

$$W = 2a + 6 = 14.48'$$

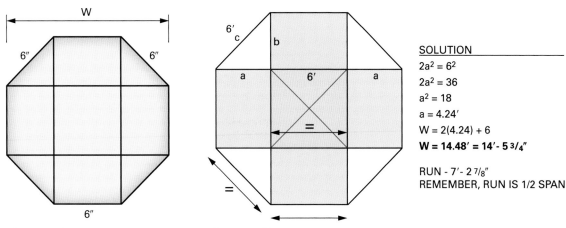

Figure 6-97 Octagon layout

Figure 6-98 Calculating the width of an octagon

SOLUTION

$2a^2 = 6^2$

$2a^2 = 36$

$a^2 = 18$

$a = 4.24'$

$W = 2(4.24) + 6$

W = 14.48' = 14'- 5 3/4"

RUN - 7'- 2 7/8"
REMEMBER, RUN IS 1/2 SPAN

How large of an octagon block is needed for 2× (1½″) rafters? See Figure 6-99.

$2a^2 = 1.5^2$; therefore $2a^2 = 2.25''$ and $a^2 = 1.125$; thus a = 1.0607″, and $1.5 + 2(1.0607) = 3.62'' = \mathbf{3\%''}$.

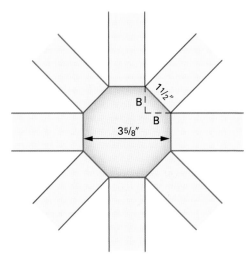

Figure 6-99 Width of an octagon block

Determining the Area of an Octagon

With "a" known, calculate the octagon's area given the length of one side. Again, there are other mathematical formulas that can shorten this process, but the

Pythagorean Theorem will be used here. Refer to Figure 6-98 for the following example:

$$6 \times 6 = \textbf{36 sq. ft.} \text{ This is the area of the center square.}$$

$$4(6 \times 4.24) = \textbf{101.76 sq. ft.} \text{ These are the four rectangles.}$$

The four small white triangles will always add up to the same area as the center square, in this case, **36 sq. ft.** Therefore,

$$\textbf{Total Area} = 36 + 101.76 + 36 = \textbf{173.76 sq. ft.}$$

Knowing the area can help to determine the amount of decking/flooring needed for construction.

Octagon Hip Unit Run

The run of an octagon hip rafter is 12.98″. This is derived below. Refer to Figure 6-96. If lines are drawn from each of the eight corners to the center, the interior angles are 360° ÷ 8 = 45°. If eight common rafters are also drawn, each will bisect the 45° angle. Therefore the angle between an octagon hip rafter and a common rafter is 22.5°. Figure 6-100 illustrates the common unit run (12″) and the 22.5° angle. By using the tangent function, it is possible to determine the octagon hip unit run.

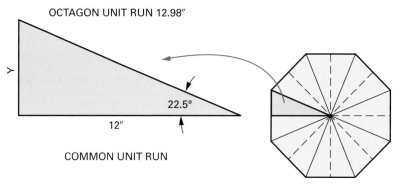

Figure 6-100 Determining octagon unit run

$$\text{Tan} = \frac{\text{opposite}}{\text{adjacent}}$$

$$\text{Tan } 22.5° = Y/_{12} \text{ (Tan } 22.5° = .4142)$$

$$.4142 \times 12 = Y; \text{ therefore, } Y = 4.97$$

Use Pythagorean Theorem to find the octagon hip unit run:

$$4.97^2 + 12^2 = c^2$$

$$c = \textbf{12.98″} = \text{octagon hip unit run}$$

For layout purposes, use **13″**. For calculations use 12.98″.

Determining Octagon Hip Line Length

Example 1: Using the Pythagorean Theorem, a unit rise of 8″ will yield an octagon hip unit length of 15.25″ (see Figure 6-101):

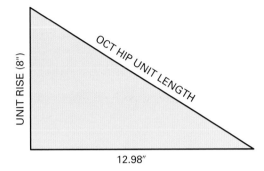

Figure 6-101 Determining octagon unit length

$$8^2 + 12.98^2 = c^2$$

$$c = 15.25″$$

Example 2: If unit rise = 4″, then the octagon hip unit length = 13.58″:

$$4^2 + 12.98^2 = c^2$$

$$c = 13.58″$$

All octagon hip unit lengths can be found using the same process.

To calculate octagon hip line length, use the same formula used to find a hip or common line length:

Octagon hip unit length $\times$ run = line length

For example, calculate a common rafter and the corresponding octagon hip rafter for an octagon structure with a 6′ side length, run = 7.24′, and slope = 4/12 (unit length 12.65).

Common rafter line length = 12.65 $\times$ 7.24 = 91.59″
(common unit length $\times$ run)

Octagon hip rafter line length = 13.58 $\times$ 7.24 = 98.32″
(octagon hip unit length $\times$ run)

Calculating Octagon Hip Jack Rafter Line Length

As it turns out, there is a constant ratio between the length of a side of an octagon and the total building width, which equals .4142. Therefore, the following is also true:

$$\text{Run of octagon jack} = \frac{\text{distance from corner}}{.4142}$$

> The line length calculated will need to be shortened depending on the method of framing at the tip.

Example Problem: The distance from the corner of an octagon to the center of the jack is 24″, and the roof slope is 6/12. How long is the jack rafter? Using the formula from above:

$$\text{Jack run} = \frac{2'}{.4142} = 4.83''$$

Therefore,

$$13.42 \times 4.83 = 64.82'' \text{ (unit length} \times \text{run)}$$

$$\text{Jack rafter} = 64.82''$$

This method assumes that the jack rafters do not attach to blocking. When blocking is used, the horizontal distance measured directly underneath the jack is used for the run.

If side ÷ width = .4142, then side = .4142 × width. Also side ÷ .4142 = width.

> *The octagon's side to width ratio can be utilized for several different calculations, providing one remembers the number ".4142."*

Octagon Hip Rafter Layout

One should thoroughly understand common rafter and hip rafter layout (see Figure 6-53) before attempting octagon hip rafter layout. The reader should be familiar enough with rafters to use previous information from this book and the instructions below to complete this task.

> *Remember, common rafters are laid out holding **12** on the body of the square with the unit rise on the tongue, and regular hip rafters are laid out using **17** on the body and the unit rise on the tongue. Octagon hip rafters are laid out using **13** on the body and the unit rise on the tongue. Octagon common rafters are the same as regular common rafters.*

- **Step 1**—Plumb lines for octagon hip rafters are laid out by holding **13** on the body of the square and the unit rise on the tongue of the square. The shortening method at the tip will depend on whether or not a center block is used in the framing. Refer back to shortening earlier in this unit (see Figures 6-94, 6-95, and 6-99).

- **Step 2**—The calculated line length (octagon hip unit length × run) is used to determine the location of the seat plumb line. This is measured from the first plumb line. The common projection can be used to determine the octagon overhang:

  Octagon hip unit length (in.) × common projection (feet)
  = octagon overhang

- **Step 3**—Octagon hip rafters either need to be dropped at the seat or the rafters have to be "backed." Follow Figure 6-53 for dropping a hip rafter, but in Step 3, instead of measuring ¾″, measure only ⁵⁄₁₆″ (.4142 × ¾″), then follow the rest of the steps.

- **Step 4**—The octagon hip rafter tail should be bevel cut at approximately 22.5° in each direction so that the fascia has a flat area to attach against (see Figure 6-54). Remember, lay out the octagon hip tail with unit rise measurement on the tongue and **13** on the body of the square.

Rafter Span Tables

There are strict guidelines to follow when determining how far a rafter can span. These are based primarily on member size, OC spacing, roof weight, and potential wind and snow load. Wind and snow loads vary regionally; thus load requirements will vary regionally, so make sure to understand the local requirements. Consult a design professional. Span tables, showing some commonly used species of wood and OC spacing variations, are located in Chapter 6 Appendix. Engineered lumber spans are not included because they vary depending on type and manufacturer. Engineered lumber spans can be obtained from manufacturers.

TRUSS CONSIDERATIONS

Types/Methods of Assembly/Bracing

There are many types of trusses, see some in Figure 6-102. Manufacturers require a specific schedule of temporary and permanent bracing, assembly recommendations, etc. These depend on many factors including, potential load, roof slope, span, type of truss, and the use and type of roofing material to be used. Truss distributors will supply a set of guidelines outlining bracing requirements and assembly procedures. Therefore those recommendations will not be included in this book. Some truss hold-down hardware is shown in Figure 6-103.

> **CAUTION**
>
> *Pay careful attention to the bracing schedule provided by the manufacturer; not all bracing can be removed after the trusses are secured. The larger the roof, the more important the bracing is.*

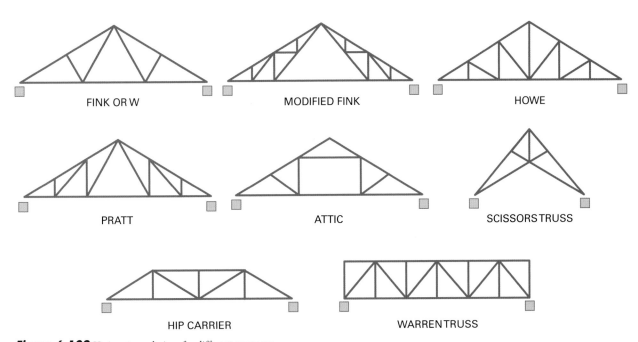

Figure 6-102 Various truss designs for different purposes

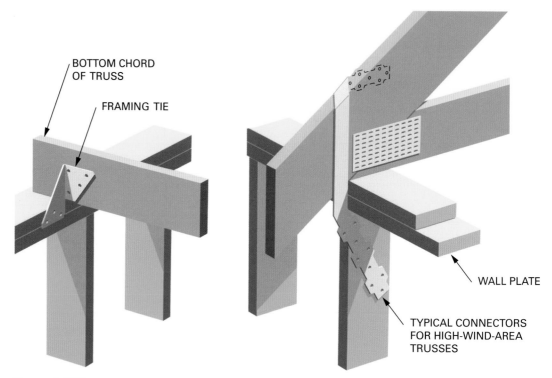

BOTTOM CHORD
OF TRUSS

FRAMING TIE

WALL PLATE

TYPICAL CONNECTORS
FOR HIGH-WIND-AREA
TRUSSES

Figure 6-103 Specialized hardware that aids in truss roof construction

Truss Uplift

A common problem encountered with trusses is a phenomenon called "uplift." This occurs when the bottom chord of a truss separates from the top plate of a partition (interior wall). Unless precautions are taken, this may cause the ceiling (sheetrock) to crack along the edge of the partition. Wood framing members gaining or losing moisture due to seasonal changes causes wood movement, which in turn triggers uplift. Wood moves (shrinks and expands) mostly across the grain and very little movement is there parallel to the grain. For example, a 10″ joist will move more in *width* than an 8′ stud will move in *length*. See Chapter 7 for more information on wood movement.

The red arrows in Figure 6-104 indicate areas prone to movement. The cumulative effect of all members shrinking or expanding can create significant movement, especially if focused in one area. If the truss is nailed tightly to an interior wall (partition) top plate, this effect may cause:

1. The wall to lift free from the sub-floor (possible);
2. The doubled plates to separate (not as likely);
3. A framing member to crack (not as likely); or
4. The truss to lift off of the top plate. This is the most common problem and generally results in a cracked sheetrock joint at the wall/ceiling intersection.

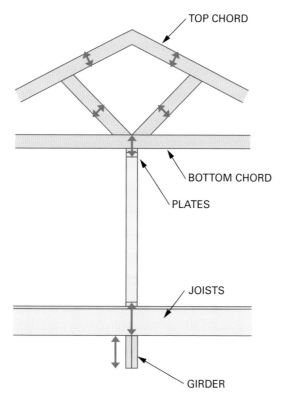

Figure 6-104 Wood is susceptible to movement primarily across the grain

Uplift is a term describing relative movement of two components— meaning the truss and its relationship to the partition. In reality, the partition sagging slightly while the truss remains stationary may contribute to "uplift" problems. However, it is still referred to as "uplift."

Figure 6-104 represents a one-story structure; however, the cumulative potential movement increases with a two-story structure.

Avoiding Uplift Damage The sheetrock joint at the truss/partition intersection should be allowed to move freely. Figures 6-105, 6-106, and 6-107 illustrate methods of allowing movement without damage. Sheetrock will tolerate seasonal movement as long as it is not stressed too quickly or over too severe of a distance. Fastening the sheetrock 16″ from the partition/truss intersection is a sufficient distance to allow the sheetrock to flex.

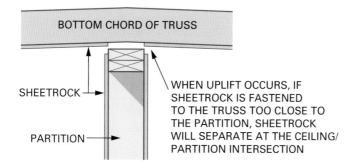

Figure 6-105 Uplift causes cracking at a truss/partition intersection

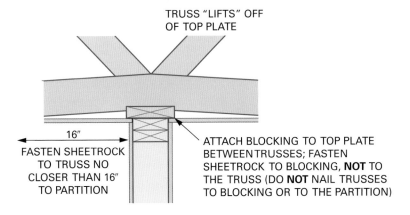

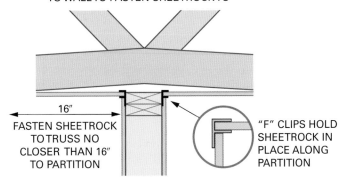

Figure 6-106 These methods allow uplift without damage

Most often the ceilings are sheetrocked before the walls; this allows the wall sheetrock to support the edges of the ceiling sheetrock. This in turn allows the first row of fasteners to be placed 16" from the partition.

Disguising Uplift Damage If there is an existing or recurring crack, it can be hidden by adding crown molding to the problem area. The crown molding should be fastened *only to the ceiling* (and not to the partition). This way, as the truss lifts seasonally, the molding will lift with the truss and hide the joint between the ceiling and the partition.

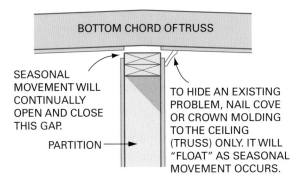

Figure 6-107 This method can disguise an existing problem

This special piece of hardware is a slotted bracket that will allow vertical movement (arrow) while simultaneously stabilizing the truss against horizontal deflection. This is an acceptable method of fastening a truss to a partition and is a good option when there is a long truss span. See Figure 6-108.

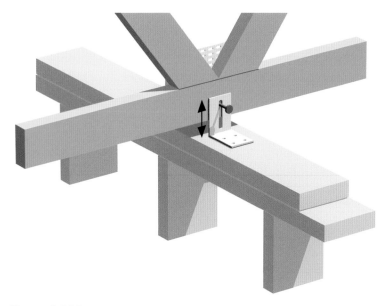

Figure 6-108 Slotted truss bracket

Advantages/Disadvantages of Trusses vs. Stick Built Roofs

Table 6-3 describes some of the issues one may want to weigh when deciding whether to frame a roof using the stick-built method versus using trusses.

Table 6-3 Stick built vs. truss built roofs

Roof Type	Advantages	Disadvantages
Stick Built Roof	• Material is often cheaper • Easier to make changes • Lumber is easy to obtain • More usable space in attic • Easier to transport lumber • No special equipment needed	• Slower to construct • More labor intensive • Size/span limitations
Truss Roof	• Faster to build • Can span longer distances • Do not need to know how to perform rafter calculations or understand rafter layout	• Some limitations to changes • Should not be cut • Materials cost higher • Need to order ahead of time • Attic trusses do not allow as much room as stick built roofs • Some problems getting trusses into difficult job sites • May need a crane

> *Some carpenters use tongue and groove roof sheathing on roofs framed 24″ OC, making ply clips unnecessary.*

Sheathing a Truss Roof

Often roof trusses are designed for 24″ OC framing (there are many exceptions); this requires the sheathing to also span this distance. In order to prevent deflection between alternate rows of sheets, panel clips are used. Sometimes these are called "H-clips" or "ply-clips." They are "H" shaped in cross section (see Figure 6-109) and they hold the sheets together at mid-span.

H-clips can also be used with stick built roofs.

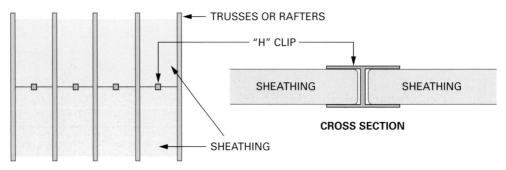

Figure 6-109 H-clips are used between trusses to keep sheathing from deflecting

CHAPTER 7

SPECIAL TOPICS

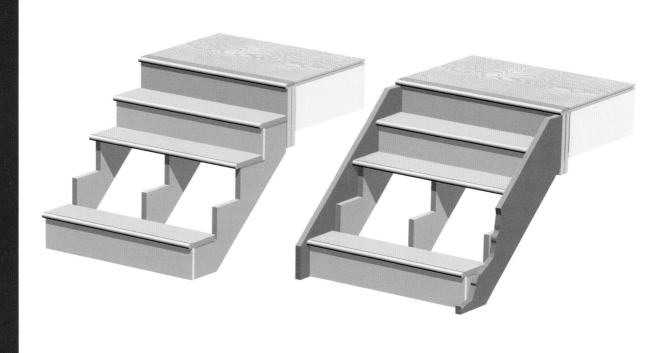

STAIRS

Since this is a book on framing, the stairs presented in this chapter will be basic and suitable for a deck, a porch, or for temporary use during construction. This is not intended as a comprehensive guide to build finished staircases.

Stair Glossary

See Figures 7-1 through 7-27 for locations of components and dimensions.

Balusters—Often people refer to these as spindles. They are spaced evenly and take up the space between the railing and the floor or tread. Codes generally require the spacing to be tight enough that a 4″ ball cannot pass between them. See Figure 7-23.

Newel post—Vertical member that supports the railings. Railings either terminate by butting into newel posts, or the railings fit over the top of them. See Figure 7-21.

Railing—Can be attached to a wall or to newel posts. Codes generally require them to be 34″–38″ high, as measured above the stair nosing, and a minimum of 36″ above landings. See Figure 7-27.

Riser—Fits perpendicular to the tread, blocks the opening between treads, and can help as a structural member.

Stringer (carriage)—The member that supports the treads and risers.

Total rise—The vertical distance from one floor level to the next.

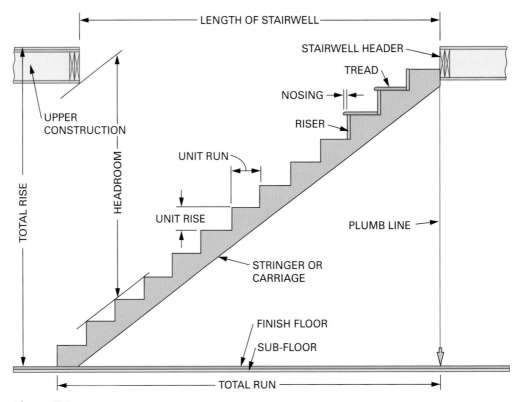

Figure 7-1 Terminology commonly used in stair construction

Total run—The horizontal distance from the face of the top riser to the face of the bottom riser. In the case where stairs turn, this is the total length found by multiplying the unit run by the number of treads.

Tread—The horizontal member that is used as a step.

Tread nosing—The amount that the tread overhangs the riser (if no riser, it is the amount the plane of one tread overlaps the one beneath it). Codes generally call for a nosing of ¾″–1¼″ and a radius of no more than ⁹⁄₁₆″. See Figure 7-2.

Unit rise—The amount of rise from the top of one tread to the top of the next. Most codes do not allow for a unit rise of more than 7¾″.

Unit run—The horizontal distance from the face of one riser to the face of the next. Most codes require a minimum of a 10″ unit run.

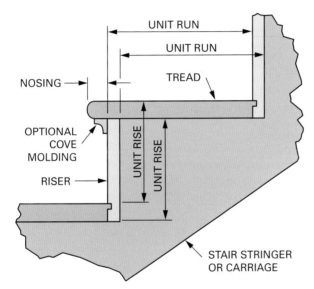

Figure 7-2 Stair terminology

Important Note: Stairs are one of the most potentially dangerous places in a home. Building codes regarding stair construction are *very* specific. Most areas use the International Residential Code therefore, those codes will be referenced in this book. Regionally different building codes are utilized and any differences, even though they may seem minor, are significant when it comes to stair construction. In some locations it is acceptable for stairs to be steeper than other areas. The reader is strongly encouraged to consult local building codes.

Below are three formulas many carpenters use to help determine the rise/run layout for stairs. These formulas do not reflect the building code and the result should always be checked for code compliance. This book will not use these formulas, but they deserve mention due to their popularity.

1. One Unit Rise plus one Unit Run should equal between 17″ and 18″.
2. Two Unit Rises plus one Unit Run should equal between 24″ and 25″.
3. One Unit Rise times one Unit Run should equal between 70″ and 75″.

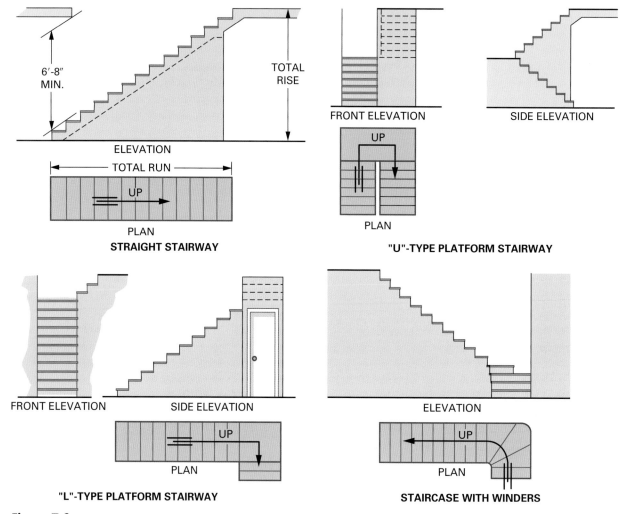

Figure 7-3 Various types of stairways

Stair Types

Staircases can be many different shapes, from straight to "L" shaped, and even circular. Figure 7-3 illustrates some common variations. No matter what the shape, the total rise and unit rise are calculated the same way.

STAIR CALCULATIONS

Determining Total Rise and Total Run

Total rise is the vertical measurement from the floor, or surface where the stairs start, to the floor where they land. In order to accurately measure total rise, it is necessary to determine the total run so the exact starting point of the staircase can be determined. This may be important if the stairs are built on sloping ground, or on a slightly sloping floor, such as in an old house (see Figure 7-4).

To determine the total run, one must multiply the number of treads in the staircase by the unit run. This may require a couple of trial-and-error calculations to find the number of treads needed.

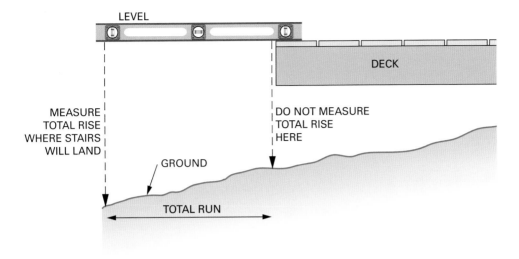

Figure 7-4 Measuring total rise

Locating the Starting Point

After determining the total run, measure horizontally outward from the upper landing the total run distance, and then plumb down to find the exact starting place of the staircase. This location is where the total rise must be measured.

Important Note: For the stair calculation and layout examples in this book, there will *always* be one less tread than risers (see Figure 7-5).

> No matter what the shape of the staircase, the total rise is measured the same way, from one finished floor to the next finished floor.

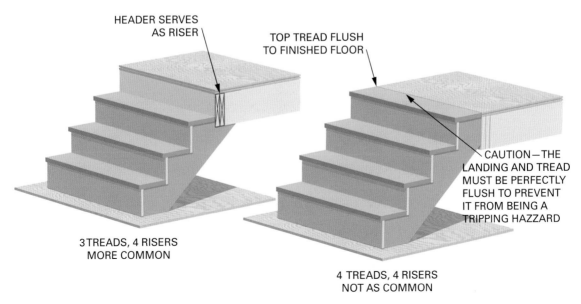

Figure 7-5 Most staircases are built like the one in the left diagram, having one less tread than risers

> For most staircases, there is one less tread than the number of risers.

NOTE

> In most cases, unit run is not the same as tread width. Most treads have a nosing that overhangs the riser. The nosing is not included in the unit run measurement (see Figure 7-2).

> Dividing by 7″ (a comfortable riser height) will yield a number that will cause a riser height of 7″ or a little more. If a riser height of less than 7″ is desired, start by dividing the total rise by 6″.

Calculating the Unit Rise

- **Step 1**—Divide the total rise by 7″. This will determine the number of risers (steps). If the number does not come out evenly, discard the decimal amount left over.
- **Step 2**—Divide the total rise by the number of risers found in Step 1. If the number does not come out evenly, convert the remaining decimal to fractions of an inch (see Chapter 2 Appendix for mathematical conversion examples). This number is the unit rise. In long staircases, it may be necessary to calculate unit rise to the nearest $\frac{1}{32}$ of an inch.

Method 1: The total rise measured is 108¾″, and the desired tread size is 11¼″ including a 1″ nosing. Unit run = 11¼″ – 1″ = 10¼″. What are the unit rise and the total run? Remember, there is one less tread than risers.

- **Step 1**—108.75″ ÷ 7 = **15.535**, therefore, 15 is the number of risers (there cannot be a fraction of a rise); discard the decimal amount for Step 2.
- **Step 2**—108.75″ ÷ 15 = 7.25 = **7¼″**. This is the unit rise.
- **Step 3**—Total run = unit run (10¼″) × number of treads (14); 10.25″ × 14 = 143.5 = 143½″.

Method 2: International Residential Code (IRC) specifies a maximum unit rise of 7¾″. Find the largest unit rise that meets IRC code for a stair with a total rise of 114½″.

- **Step 1**—Determine the number of risers: 114.5″ ÷ 7.75″ = 14.77. Since a stair cannot be built with a partial rise, round up 14.77 to 15.
- **Step 2**—114.5″ ÷ 15 = 7.633″ = 7⅝″ unit rise.

STAIR LAYOUT

Stair layout can be performed when the number of treads, desired unit rise, and unit run are known. This process is much easier and more accurate if a framing square and a pair of stair gauges are used. See Figures 7-6 through 7-11.

- **Step 1**—Attach the stair gauges to the square so that when the square is placed against the stringer (board that is being laid out), the unit rise and unit run numbers align with the edge of the board. See Figure 7-6.
- **Step 2**—Place the square a few inches up from the end of the stringer as shown in Figure 7-7, and trace along the bottom edges of the square the unit rise and unit run.
- **Step 3**—Slide the square along the board and repeat for as many treads as are planned. See Figure 7-8

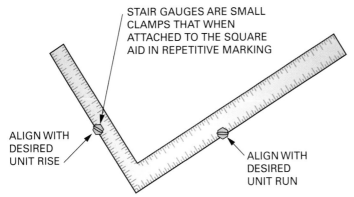

STAIR GAUGES ARE SMALL CLAMPS THAT WHEN ATTACHED TO THE SQUARE AID IN REPETITIVE MARKING

ALIGN WITH DESIRED UNIT RISE

ALIGN WITH DESIRED UNIT RUN

NOTE:
WHEN USING STAIR GAUGES, ALIGN THE UNIT RISE AND UNIT RUN ALONG THE EDGE OF THE STRINGER, THEN PLACE THE GAUGES. DUE TO THEIR SHAPE, THE GAUGES WILL NOT ALIGN EXACTLY ON THE UNIT RISE AND UNIT RUN NUMBERS SHOWN ON THE SQUARE. HOWEVER, THE LAYOUT MARKS WILL BE ACCURATE.

Figure 7-6 Laying out a stair stringer

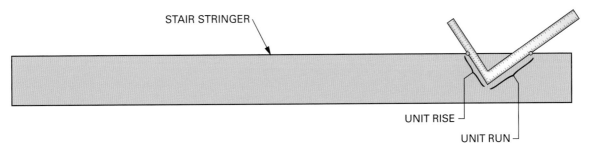

STAIR STRINGER

UNIT RISE

UNIT RUN

Figure 7-7 Step 2

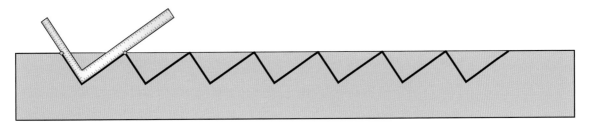

Figure 7-8 Step 3

- **Step 4**—Square down from the first (bottom) tread drawn and draw the bottom riser (Figure 7-9). *The height of the bottom riser equals the unit rise minus the tread thickness.* This will keep the bottom rise from being too high.

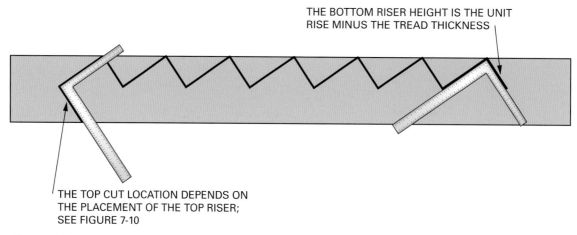

THE BOTTOM RISER HEIGHT IS THE UNIT
RISE MINUS THE TREAD THICKNESS

THE TOP CUT LOCATION DEPENDS ON
THE PLACEMENT OF THE TOP RISER;
SEE FIGURE 7-10

Figure 7-9 Steps 4 and 5

- **Step 5**—At the last (top) tread, square down from the back of the tread toward the bottom of the stringer; this will represent the back edge of the stringer where it abuts the top riser. Often the top riser is attached to the face of the stairwell header (see Figures 7-1 and 7-10).

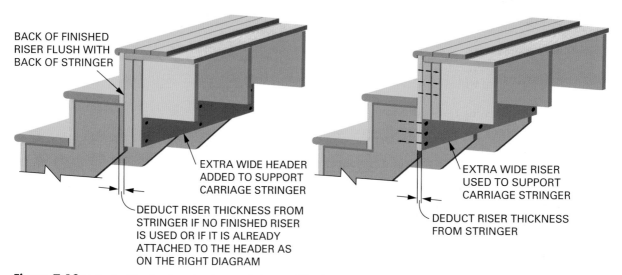

BACK OF FINISHED
RISER FLUSH WITH
BACK OF STRINGER

EXTRA WIDE HEADER
ADDED TO SUPPORT
CARRIAGE STRINGER

DEDUCT RISER THICKNESS FROM
STRINGER IF NO FINISHED RISER
IS USED OR IF IT IS ALREADY
ATTACHED TO THE HEADER AS
ON THE RIGHT DIAGRAM

EXTRA WIDE RISER
USED TO SUPPORT
CARRIAGE STRINGER

DEDUCT RISER THICKNESS
FROM STRINGER

Figure 7-10 Methods of framing the stair stringer to their stairwell header

- **Step 6**—Cut out the stringer along the lines drawn and use it to trace as many more stringers as are needed. Wide stairs may need one or more intermediate stringers (Figure 7-20). By tracing, all stringers will be identical.

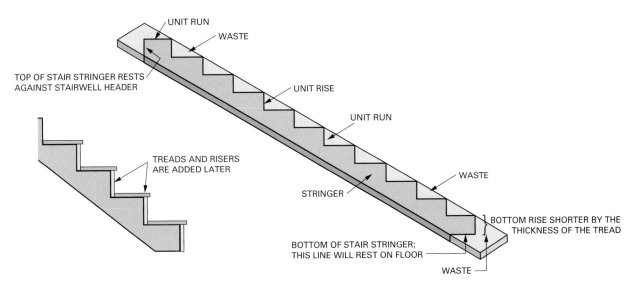

Figure 7-11 A completed stair stringer layout

MULTIPLE SOLUTIONS

Occasionally, it is necessary to build stairs to the limits of the building code. This is not the best practice but space may require it. For example, if a staircase were too long (total run), and there was not enough room between the bottom stair and a wall (codes generally require 3′ or more of space at the bottom of a stair). There are two ways to reduce the total run: either shorten the unit run or decrease the number of treads, which will also shorten the total run. See Figure 7-12.

- **Method 1, Shortening the Unit Run**—A carpenter plans a 14-tread staircase using 11¼″ treads with a ¾″ nosing (a 10½″ unit run). If the unit run was shortened to 10″, then for every tread, the staircase's total run will be shortened by ½″. 14 treads × ½″ per tread = 7″ total run shorter!
- **Method 2, Dropping a Tread**—A staircase has a total rise of 105″. The plan is to have a unit run of 10″ and 15 risers at 7″ per rise, so that 7″ × 15 = 105″ total rise. If the staircase were built with 14 risers instead of 15, then 105″ ÷ 14 = 7.5″ unit rise. The unit rise increases to 7.5″; however, the staircase will have one less tread, which translates to one less unit of run. This will allow the staircase to take up 10″ less total run.

By using both of these methods on the same staircase, a total of 17″ can be taken off of the total run of the staircase.

In some cases it may be possible to drop two treads. This in combination with a shortened unit run gives the carpenter some versatility in sizing the total run of the staircase.

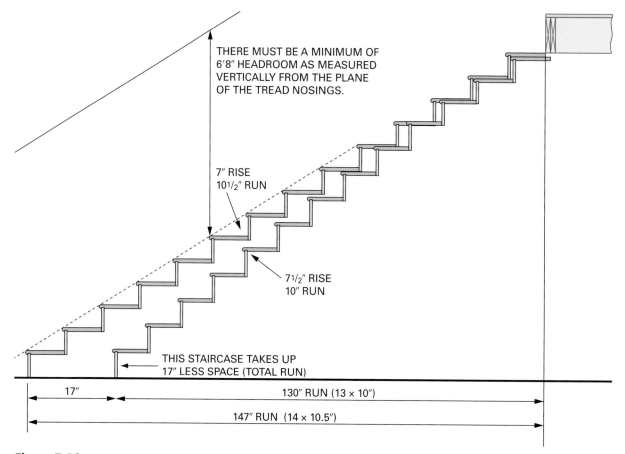

THERE MUST BE A MINIMUM OF 6'8" HEADROOM AS MEASURED VERTICALLY FROM THE PLANE OF THE TREAD NOSINGS.

7" RISE
10½" RUN

7½" RISE
10" RUN

THIS STAIRCASE TAKES UP 17" LESS SPACE (TOTAL RUN)

17"

130" RUN (13 × 10")

147" RUN (14 × 10.5")

Figure 7-12 Often there are two or more possible solutions for a staircase

Using the above methods, staircases can also be designed based on the available total run. Just divide the total run by the number of treads to obtain the maximum unit run. Make sure the final figures are code compliant before starting construction.

Building Stairs with a Landing (Platform)

Often stairs need to have an intermediate landing or "platform." Calculating unit rise is done the same way as previously. The platform takes the place of one tread and riser.

It is important to maintain the same unit rise spacing on the flights below and above the platform; therefore, the height and location of the platform *must* be planned into the overall design (Figure 7-13).

Methods of Construction

Some of the basic methods of stair construction are illustrated below. See Figures 7-14 through 7-18.

Cleated stairs (see Figure 7-18) will be dangerous if the cleats are not of substantial size and strength and if not fastened securely. Make sure cleats are not split or cracked. Cleated stairs may not be allowed in some areas.

> *Building codes only allow ⅜" maximum variation between the largest rise and the smallest rise in a staircase. Generally, if there is such a variation, it is on either the first or last tread (this can be due to a change in the type of flooring used).*

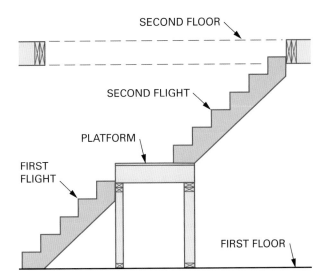

Figure 7-13 The platform height and location are carefully planned into the stair design so that *all* unit rises and unit runs within the staircase are consistent

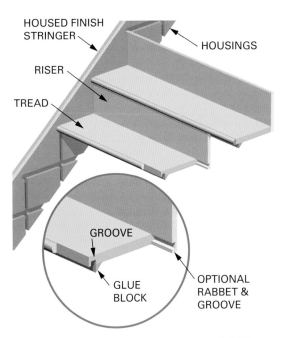

HOUSED FINISHED STRINGER METHOD

Figure 7-14 With the use of a router and a special jig, a housed stringer can be constructed: the dadoed stringer creates a finished appearance with the treads and risers

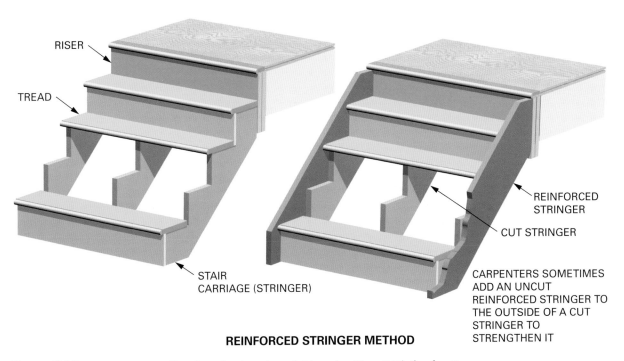

REINFORCED STRINGER METHOD

Figure 7-15 A cut stringer is significantly weaker than a housed stringer (see Figure 7-14); therefore, it may be necessary to reinforce it or construct it with engineered lumber such as an LVL beam

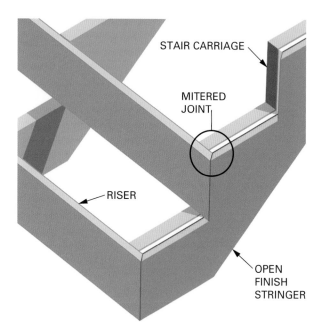

Figure 7-16 A mitered joint is made between the risers and open finish stringer so no end grain is exposed, giving a finished appearance

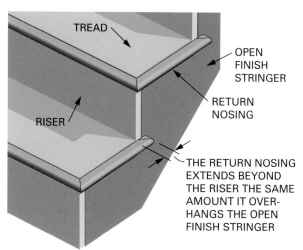

Figure 7-17 A return nosing is mitered to the open end of the tread, finishing the end of the tread

CAUTION

Cleated stairs will be dangerous if the cleats are not of substantial size or strength and if not fastened properly. Cleated stairs may not be allowed in some areas.

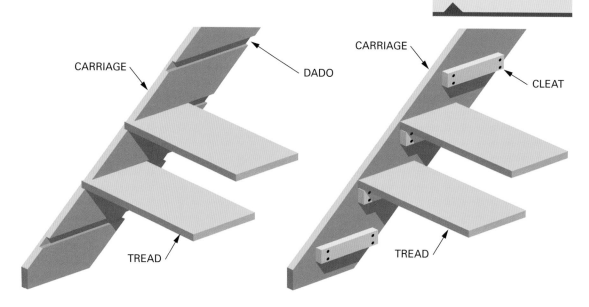

NOTE: THE OPEN RISERS (LACK OF RISERS) MAY BE A VIOLATION OF THE BUILDING CODE UNLESS THE OPENING IS SMALL ENOUGH THAT 4" BALL CANNOT PASS BETWEEN THE TREADS.

Figure 7-18 The dadoed method (left) or the cleated method may also be used

Assembly

Stairs can be fully assembled off-site and then brought in as a unit and placed, or they can be built in place. The following is a suggested method of assembling a set of stairs complete with risers:

- **Step 1**—Space the stringers evenly (if more than two are used); a couple of temporary cleats may be helpful to holding the spacing.
- **Step 2**—Place the bottom two risers first, followed by the bottom tread.
- **Step 3**—Alternate placement of riser, tread, riser, tread, working from the bottom to top of the staircase; this way it is easy to attach the risers to the treads by fastening through the backs of the risers.

When building rough stairs, it is acceptable to secure the treads and risers to the stringer through their faces. Screws are recommended because they are less likely to come loose and create a trip hazard.

If the stairs have been assembled off-site, and will be opened on one or both sides (not in between walls), then the base must be fastened to the floor (see Figure 7-19). Otherwise, the stringers can be fastened to the walls and/or supported by a wall underneath the stringer (see Figure 7-20). Fastening tops of stringers is often planned into the construction of the landing, as in Figure 7-10.

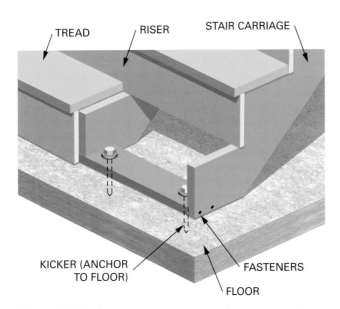

Figure 7-19 A kicker plate may be used to anchor a carriage to floor

Newel Posts

There are many codes associated with stair railings/balustrade systems, including: height ranges, baluster spacing, profiles, placement in relationship to the wall, and more. Therefore, placement of the balustrade systems must be *very* precise. Some find this more difficult than building the stairs.

Newel posts are generally placed with the center of the newel flush with the face of the risers (see Figure 7-21) *and* the center aligned with the outside of the stringer. Newel posts need to be plumbed in both directions and secured **firmly**.

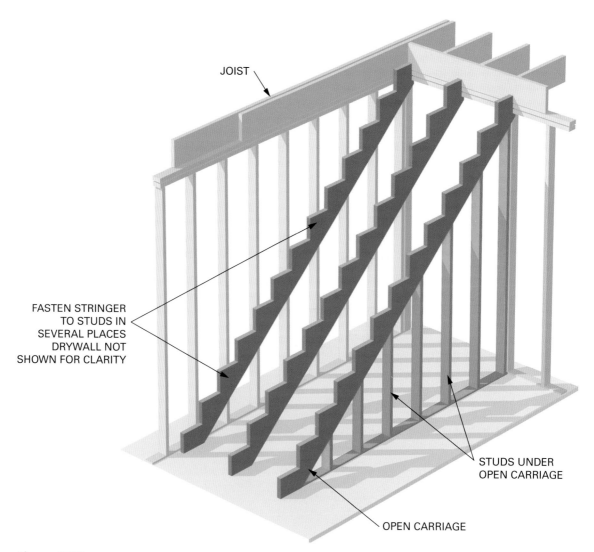

JOIST

FASTEN STRINGER
TO STUDS IN
SEVERAL PLACES
DRYWALL NOT
SHOWN FOR CLARITY

STUDS UNDER
OPEN CARRIAGE

OPEN CARRIAGE

Figure 7-20 One stringer is fastened to the partition and another is supported by a wall

RAILINGS/BALUSTRADE SYSTEMS

After the newel posts have been secured, the railings can be cut and attached. An easy method of marking and cutting a stair railing to fit to newel posts is illustrated in Figures 7-21 through 7-27.

Most building codes require the railings to be placed between 34″–38″ as measured vertically above the tread nosing. The slope of the rail *must* run parallel to the slope of the stairs (in other words, height **cannot** be 34″ above the bottom nosing and 38″ above the top nosing). Railings also have specific profile requirements that allow them to be "grasped." As always, check your local building codes.

Railings can be secured to the newel posts with lag bolts or large screws; holes are plugged to hide the heads.

Railings can also be attached with special hidden hardware called rail bolts. Rail bolts are most commonly used on finished balustrades indoors (see Figure 7-26).

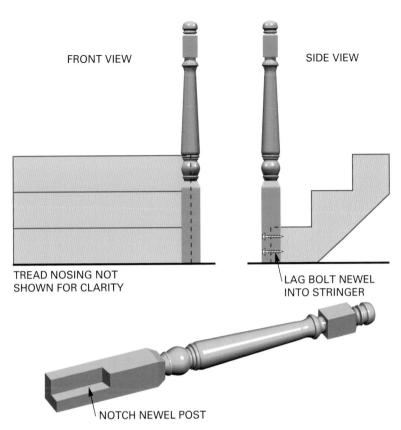

FRONT VIEW

SIDE VIEW

TREAD NOSING NOT
SHOWN FOR CLARITY

LAG BOLT NEWEL
INTO STRINGER

Always secure newel posts very firmly. They are required to withstand strong forces. Whenever possible, make them long enough to secure them to the joists, to the stringer, or, in some cases (bottom of stairs on decks), bury the post base in the ground.

NOTCH NEWEL POST

Figure 7-21 The starting newel is notched to fit over the first step

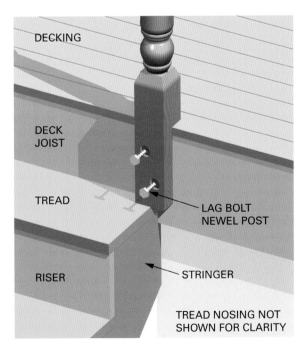

DECKING

DECK
JOIST

TREAD

LAG BOLT
NEWEL POST

RISER

STRINGER

TREAD NOSING NOT
SHOWN FOR CLARITY

Figure 7-22 The center of the upper newel post is centered on the edge of the deck joist

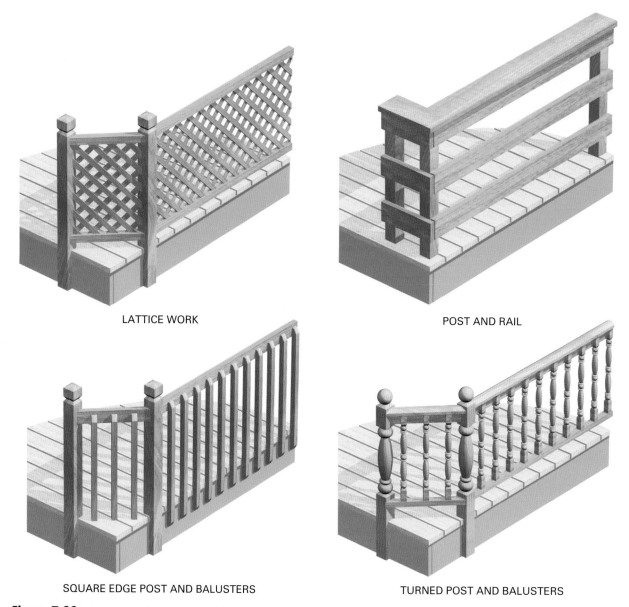

LATTICE WORK

POST AND RAIL

SQUARE EDGE POST AND BALUSTERS

TURNED POST AND BALUSTERS

Figure 7-23 Balustrades can be constructed with various designs

Balusters

Many different sizes and profiles are available. Codes generally require spacing so that a 4″ diameter ball cannot pass between balusters. Balusters can either be placed between a top and bottom railing or between the top railing and the tread/floor.

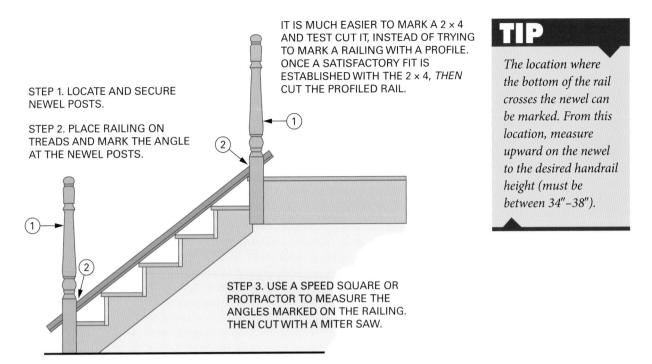

STEP 1. LOCATE AND SECURE NEWEL POSTS.

STEP 2. PLACE RAILING ON TREADS AND MARK THE ANGLE AT THE NEWEL POSTS.

IT IS MUCH EASIER TO MARK A 2 × 4 AND TEST CUT IT, INSTEAD OF TRYING TO MARK A RAILING WITH A PROFILE. ONCE A SATISFACTORY FIT IS ESTABLISHED WITH THE 2 × 4, *THEN* CUT THE PROFILED RAIL.

STEP 3. USE A SPEED SQUARE OR PROTRACTOR TO MEASURE THE ANGLES MARKED ON THE RAILING. THEN CUT WITH A MITER SAW.

TIP

The location where the bottom of the rail crosses the newel can be marked. From this location, measure upward on the newel to the desired handrail height (must be between 34″–38″).

Figure 7-24 Marking and cutting railings

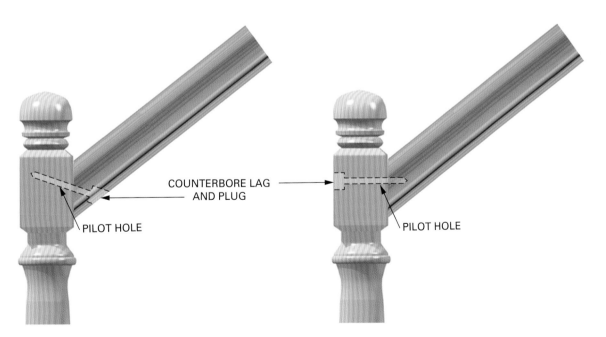

COUNTERBORE LAG AND PLUG

PILOT HOLE

PILOT HOLE

Figure 7-25 This handrail is fastened to newel posts with lag bolts. The use of nails when constructing balustrades is discouraged

After all holes are drilled, secure the lag side of the rail bolt to the newell. Next, place the rail over the machine threaded end of the bolt, carefully reach up through the 1" hole, and place the nut on the end of the rail bolt. It can be tightened with a screwdriver by turning the special fins on the bolt.

Baluster spacing is regulated by building codes. A 4" ball cannot pass between the widest opening in the balusters. Although not a code issue, baluster spacing should be consistent in order to look attractive.

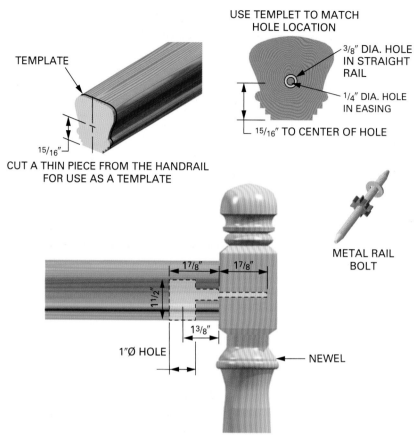

USE TEMPLET TO MATCH HOLE LOCATION

TEMPLATE

3/8" DIA. HOLE IN STRAIGHT RAIL

1/4" DIA. HOLE IN EASING

15/16" TO CENTER OF HOLE

15/16"

CUT A THIN PIECE FROM THE HANDRAIL FOR USE AS A TEMPLATE

METAL RAIL BOLT

1 7/8"   1 7/8"

1 1/2"

1 3/8"

1"Ø HOLE

NEWEL

Figure 7-26 A template is made to mark the ends of handrails and fittings for joining with rail bolts. The ends are drilled to specific depths and diameters

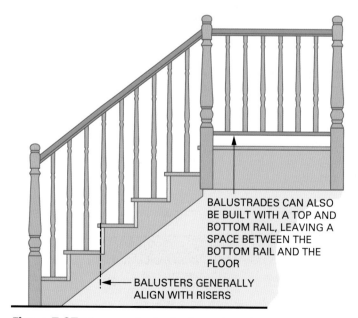

BALUSTRADES CAN ALSO BE BUILT WITH A TOP AND BOTTOM RAIL, LEAVING A SPACE BETWEEN THE BOTTOM RAIL AND THE FLOOR

BALUSTERS GENERALLY ALIGN WITH RISERS

Figure 7-27 Balusters can be placed with or without a bottom rail

Baluster layout Balustrades are designed to be visually pleasing. Generally, the spaces between the last balusters, on either end of a section, are identical. This uniformity is easy to achieve when using a story pole.

Baluster spacing options are limited due to the building code restrictions; therefore, layout of baluster spacing is important. A suggested method is to layout a story pole to aid this process. Figure 7-28 represents a stick (it can be any straight piece of lumber, a 1 × 3, 2 × 4, scrap plywood, etc.). The stick has marks (brown squares) placed on it to indicate balusters; white indicates the spacing between the balusters. The black dashed lines represent the center of the story pole. For example, the balusters may be 2″ × 2″ and the space between the balusters 3¾″.

Start by making a story pole, based on the spacing and baluster size. Only one story pole is necessary, the two centerlines can be on the same pole even though one of them will not be in the exact center (dashed line). Two story poles are shown in Figure 7-28 to illustrate this concept.

Figure 7-28 A story pole is helpful when laying out balustrades

Next, cut the railings to the exact length and mark the center of the railing.

Place the story pole next to the railing aligning the centerline of the story pole with the centerline of the railing. Complete the layout by transferring the marks from the story pole to the railing.

Notice that one-story pole shows the center of a baluster aligning with the center of the rail, while the other shows the center of a *space* between balusters aligning with the center of the railing. It only takes a second to try both ways to see which layout will work best.

Using the story pole has the added benefit of saving time, especially when building multiple sections of railings. All one has to do is place the story pole next to the railing, decide which layout will look best, then transfer the marks. In Figure 7-29, it is readily apparent that the upper layout will not work because there will be half of a baluster at each end of the balustrade. The lower option will work much better.

If building a sloped balustrade for stairs, the slope will have to be considered for layout spacing.

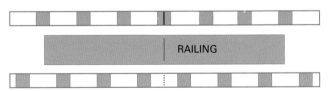

Figure 7-29 Comparing two possible layouts using a story pole shows the upper layout will not work in this case

PORCH/DECK FRAMING

Framing the floor of a deck or porch utilizes many of the concepts discussed in the earlier chapters. Therefore, this unit will build on previous information.

Porches and decks are exposed to the weather and must be constructed to maximize longevity. Below are several suggestions. See Figures 7-30 and 7-31.

1. Use decay-resistant lumber. Support posts, joists, and girders should be pressure treated, or another code-compliant decay-resistant material.

2. Decking material, balustrade systems, and other wood should also be decay resistant. Wood including redwood, cypress, cedar, white oak, and certain others are somewhat resistant to decay. Heartwood (the center of the log) is generally better than sapwood (the outer log); some of the wood mentioned, such as cedar, have many varieties, some of which are better than others. There are also many different types of engineered products that work well.

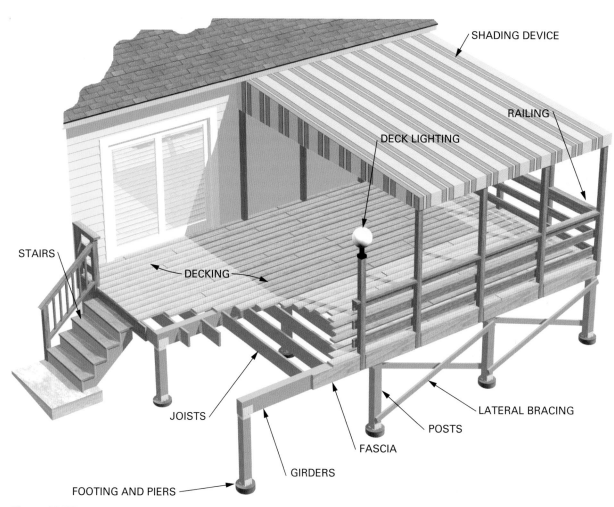

Figure 7-30 The components of a deck

3. If using untreated wood, coat all surfaces with preservative, including any cut ends.

4. Slope flat surfaces (such as railings) to shed water.

5. Slope the ground away from the deck so that water cannot collect beneath it and cause excessive humidity.

6. VENT, VENT, VENT!!! Do not allow air to be trapped beneath a deck or stairs or inside hollow cavities such as newel posts and columns. Make sure there are entry and exit points to encourage air flow.

7. If a solid floor (no gaps) such as tongue-and-groove flooring is used on a porch, the floor should slope slightly to allow for drainage and it should overhang the rim joist slightly.

8. If building a covered porch, allow the rafters to overhang the porch by a few inches; this will help to channel water away.

> *Some places with strong environmental concerns, such as areas near lakes, may not permit use of pressure-treated lumber. Check local codes.*

> *Any small entry and exit vents will need screening placed over them to keep unwanted nest-building insects out.*

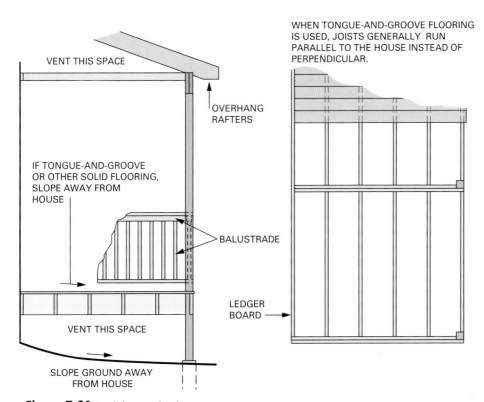

Figure 7-31 Porch framing details

Deck/Porch Construction

Posts are located based on spacing necessary to carry the load. This may vary depending on the size of joists used. Batter boards (see Figure 7-32) can be used during layout to help with accuracy of post placement.

Batter boards consist of simple horizontal boards where string lines can be attached and run from one batter board to another. At the intersections of the strings, a plumb bob (or level) can be used to mark specific locations. These boards are used during initial layout and then the strings are removed for the excavation process. After excavation, the strings are replaced to re-establish layout locations. They are most commonly used to help locate footings, foundation walls, and posts.

FOOTING AND POST LAYOUT AND EXCAVATION

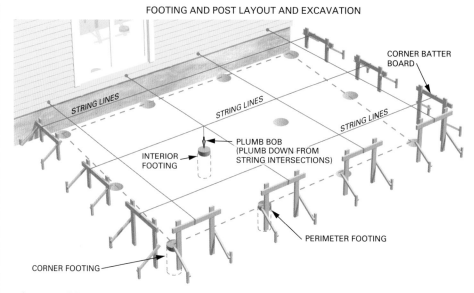

Figure 7-32 Footing and post layout and excavation

After holes have been excavated, concrete footings are placed, followed by the posts. Accurate placement of the footing is important. The post should be centered on the footing. Using batter boards and string lines can aid in accurate post placement.

Figure 7-33 shows a post placed on a footing below ground level and secured by pouring concrete around the post.

Footing depth and size will depend on frost line, soil type, and potential load.

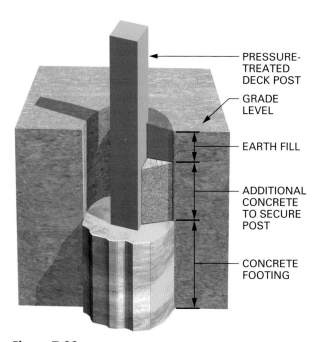

Figure 7-33 Treated post placed on a concrete footing

After posts are placed, they are cut off to their necessary heights. Joists, or girders if needed, are fastened to the posts. The use of girders will depend on the size of the deck. See Figure 7-34.

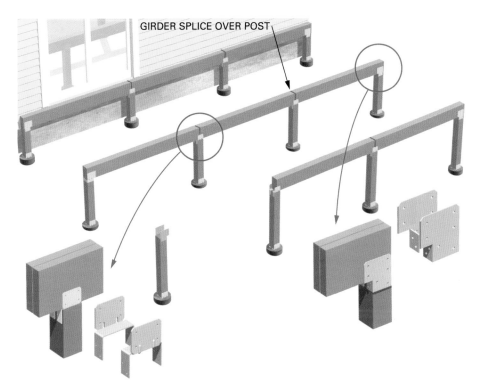

GIRDER SPLICE OVER POST

Figure 7-34 Girders are installed with their crowned edges up and anchored to supporting posts, sometimes with special hardware

A ledger board is attached to the house to support one end of the joists.

Joists are placed after the ledger and girders have been secured. Spacing of joists depends on span, load, and decking type. See Figures 7-35 and 7-36.

Next, newel posts can be attached and then decking added. As with stairs, discussed earlier in the chapter, newel posts must be able to withstand strong lateral forces. A recommended method of securing newel posts is to make sure they are long enough to fasten to the framing beneath the deck. Fasten with carriage bolts or lag bolts. See Figure 7-37.

Special post anchors are available; they can be placed in the concrete before it has set, or fastened to the concrete after it has hardened. Some carpenters prefer to use round tubular forms for the concrete and to place the bottom of the posts above ground level.

CAUTION

Ledger boards fastened to the house are subject to large forces and must be fastened according to building code requirements. Lags are often used to fasten the ledger board, with special attention to size and OC spacing of the lag bolts.

When hardware and fasteners are used outdoors, make sure they are made of material approved for outdoor purposes, such as stainless steel or galvanized.

Depending on several factors including deck height above ground, space needed beneath the deck, framing style, or a carpenter's preferences of construction methods, the girders may be framed beneath the joists instead of framed flush, as shown in the accompanying diagrams.

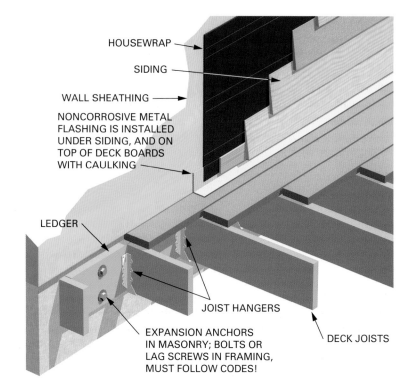

HOUSEWRAP

SIDING

WALL SHEATHING

NONCORROSIVE METAL FLASHING IS INSTALLED UNDER SIDING, AND ON TOP OF DECK BOARDS WITH CAULKING

LEDGER

JOIST HANGERS

EXPANSION ANCHORS IN MASONRY; BOLTS OR LAG SCREWS IN FRAMING, MUST FOLLOW CODES!

DECK JOISTS

Figure 7-35 A ledger is made weather tight with flashing

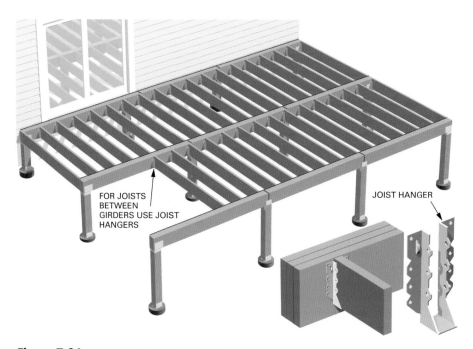

FOR JOISTS BETWEEN GIRDERS USE JOIST HANGERS

JOIST HANGER

Figure 7-36 Joists being placed between girders

A BRIEF OVERVIEW OF WOOD MOVEMENT

There is an age-old debate concerning the orientation of deck boards. Should they be placed bark side up or down? Not the actual bark, but the side of the board closest to the bark. See Figure 7-39.

Instead of settling this debate, which will not happen, one has to understand *and be able to anticipate* how wood will move under different conditions, and then make an informed decision.

The answer to the question is: one should always use their best judgment based upon personal knowledge of local environmental conditions, the moisture content of the wood being used, and the appearance of the wood.

As wood dries it shrinks, but not until it drops below approximately 30% moisture content (MC). This percentage varies slightly with species. As it dries from 30% downward toward 0%, it will move (shrink) considerably.

Heartwood, nearer the center of the tree, is more stable than sapwood, nearer the bark, and heartwood will shrink slightly less than sapwood, as it loses moisture. Because the sapwood shrinks more than heartwood (see Figure 7-41), it will have a tendency to cup slightly *away* from the center of the tree as it dries. Furthermore, as wood dries, the tangential grain of the wood shrinks more than the radial grain (see Figure 7-40). The longitudinal grain (length of a board) will shrink a negligible amount.

Using the information above, study Figure 7-41 to understand which way different pieces of wood will move as they *lose* moisture.

> Some carpenters use a special H-shaped piece of hardware that spaces the ledger board away from the house.

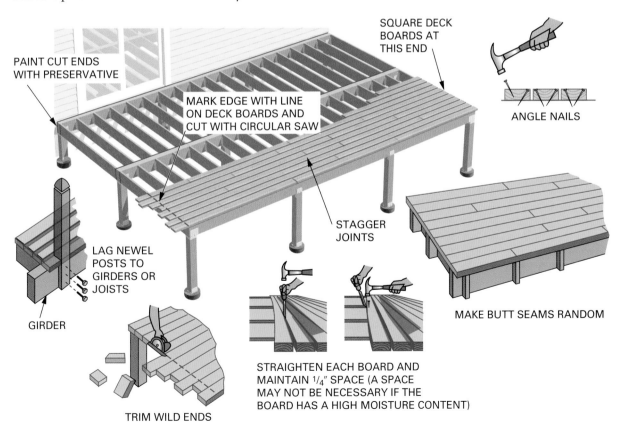

SQUARE DECK BOARDS AT THIS END

ANGLE NAILS

PAINT CUT ENDS WITH PRESERVATIVE

MARK EDGE WITH LINE ON DECK BOARDS AND CUT WITH CIRCULAR SAW

STAGGER JOINTS

LAG NEWEL POSTS TO GIRDERS OR JOISTS

GIRDER

TRIM WILD ENDS

STRAIGHTEN EACH BOARD AND MAINTAIN 1/4″ SPACE (A SPACE MAY NOT BE NECESSARY IF THE BOARD HAS A HIGH MOISTURE CONTENT)

MAKE BUTT SEAMS RANDOM

Figure 7-37 Techniques for installing deck boards

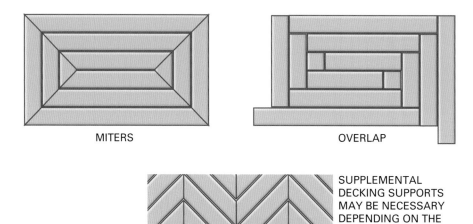

MITERS

OVERLAP

SUPPLEMENTAL DECKING SUPPORTS MAY BE NECESSARY DEPENDING ON THE DECKING PATTERN

HERRINGBONE

Moisture Content (MC)—*This is expressed as a percentage of the water weight in the wood. For example, a 100% MC means there is as much water weight as wood weight in the sample.*

Figure 7-38 Deck boards can be installed with various patterns

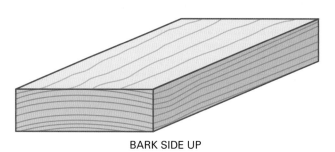

BARK SIDE UP

Pressure-treated wood is often delivered to the jobsite with an extremely high MC (40% +). Because the MC is so high, the wood has to drop many percentage points of its water weight before it starts to shrink. This is why it may look good for several days or even weeks after installation, before showing signs of shrinking.

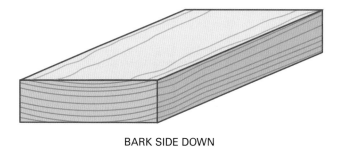

BARK SIDE DOWN

Figure 7-39 Deck board orientation: Which way up?

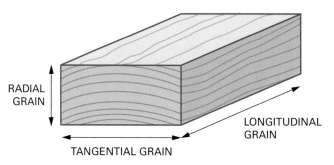

RADIAL GRAIN

TANGENTIAL GRAIN

LONGITUDINAL GRAIN

Figure 7-40 Wood grain types

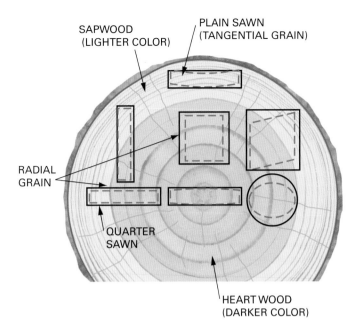

SAPWOOD
(LIGHTER COLOR)

PLAIN SAWN
(TANGENTIAL GRAIN)

RADIAL
GRAIN

QUARTER
SAWN

HEART WOOD
(DARKER COLOR)

NOTE: THE DEFORMATION SHOWN IN RED HAS BEEN SLIGHTLY EXAGGERATED.

Figure 7-41 As lumber dries, the annular rings become shorter, sometimes causing wood to deform

Wood that is quarter sawn has a radial grain (having the growth rings running nearly perpendicular to the face) and is more stable than plain-sawn lumber. See Figure 7-42. However, quarter-sawn boards are more expensive and not as readily available.

Figure 7-42 A plain-sawn board, as it loses moisture, will tend to cup away from the center of the tree

Deck boards should be placed, whenever possible, so the *cupped side is down* or will be down as the board adjusts to local humidity. This will help the boards to shed water instead of letting it pool. Furthermore, there will be less of a tripping hazard. See Figures 7-43 and 7-44.

Figure 7-43 Deck boards should be placed to cup down

This does not mean that deck boards should always be placed bark side down! Nor does it mean the bark side should always be placed up! Make a decision based on local environmental conditions, the moisture content of the wood being placed, and the condition of the wood being placed, and the look of each individual piece.

CAUTION

When placing deck boards that have not adjusted to local moisture (humidity) conditions, remember that as wood gains or loses moisture, it moves primarily across the grain. It will shrink or grow in width if moisture is released or added. If the wood is firmly secured and has high moisture content, and then it loses moisture, it will not be able to move, because of the fasteners holding it in place, and it will check and possibly split! See Figure 7-48 for a solution.

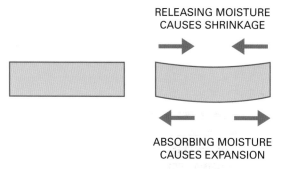

Figure 7-44 Excessive moisture *beneath* any board can cause upward cupping

The top surfaces of decks and porches get sunlight, which has a drying effect. The areas underneath porches and decks are often damp and not well vented. A situation like this will nearly always cause the bottoms of boards to retain moisture while the top surfaces lose moisture, causing upward cupping, regardless of the boards' orientation.

Solutions are offered below, but read on to discover more about movement issues.

Figures 7-45 through 7-47 illustrate the problems that can occur with a mitered floor.

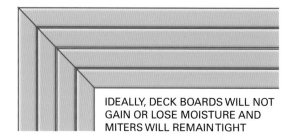

Figure 7-45 Boards have not gained or lost moisture

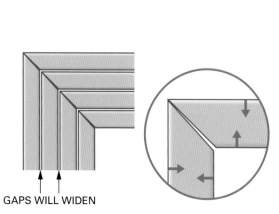

Figure 7-46 Boards have lost moisture after being placed

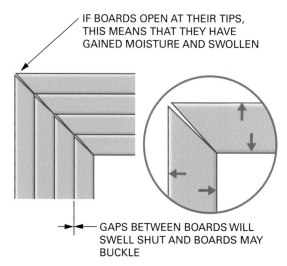

Figure 7-47 Boards have gained moisture after being placed

Important Note: Wood will gain and lose moisture as seasons change. Therefore, minor fluctuations will occur throughout the year. "Sealing" the lumber by coating it with paint, stain, or water sealer may slow the process but it will *not* stop it!

Solutions

- Pressure-treated lumber is often sold with a moisture content of more than 40%; sometimes it is as high as 100% or more. To reduce movement, it should be acclimated to the surrounding area to equalize with the local humidity until it reaches what is called "Equilibrium Moisture Content" (EMC). Once wood has reached EMC, it will shrink and expand only slightly as humidity changes and will generally not cause significant movement problems. This drying process or equalization process can take several weeks, or months. If the wood is fastened while wet, the deck boards can be placed tightly together; as they shrink, gaps will form.

- Another solution is to fasten only one side of each board so that it can shrink without the risk of splitting. After it dries for several weeks, or even months, the other side can be secured (it is possible that some warping may occur). This technique will not help with miters. See Figure 7-48.

> *Wood with a MC of over 100% is possible. 100% MC means that half of the board's weight is due to moisture.*

ALLOWING BOARDS TO SHRINK WITHOUT SPLITTING

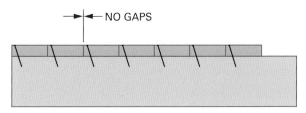

1. PLACE DECK BOARDS TIGHTLY TOGETHER AND SECURE ONLY ONE SIDE OF THEM.

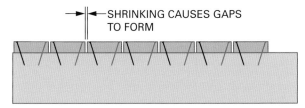

2. AFTER DECK BOARDS HAVE SHRUNK AND GAPS HAVE FORMED, SECURE THE OTHER SIDE OF THE BOARDS (RED FASTENERS). BOARDS WILL CONTINUE TO MOVE BUT THE FUTURE MOVEMENT WILL BE MINIMAL.

Figure 7-48 Boards will check or split if fastened when they have a high MC

- If waiting months for wood to dry is out of the question, consider purchasing dry wood. In this case it is possible the wood may be too dry and needs to acclimate by being allowed to *gain* moisture before installing it outdoors

*Moisture meters that test moisture content of wood are relatively common. Use a moisture meter to test the MC of wood (such as an exterior part of the house), and then test the wood being purchased. This can take the guesswork out of the process. If wood is at EMC before installation, it will gain or lose only a small percentage of moisture throughout the year, and there will be **fewer problems** associated with wood movement.*

(choices of wood type may be limited). If gaps between dry deck boards are desired, the wood will have to be purposely placed with gaps between boards.

- The area beneath the deck/porch should be well vented, dry, and drained; this may help prevent a moisture gain on the back of the boards.

- Use a composite decking material. It does not tend to shrink or expand and is generally more expensive than wood, but may be worth the extra cost.

- An alternative solution to using miters at the corners is to use an overlapping pattern (see Figure 7-49). This method is decorative but does not have miters to contend with. Miters can be attractive, but they can also look unsightly when dealing with fluctuating MC.

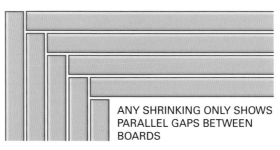

ANY SHRINKING ONLY SHOWS PARALLEL GAPS BETWEEN BOARDS

Figure 7-49 Overlapping pattern

Here are some general rules of thumb for placing wood deck boards (not tongue-and-groove-type decking):

- If possible let the wood reach EMC before placing it.
- **If wood is placed with a high MC:**
  - If it is already cupped, place cupped side down regardless of grain.
  - If wood is flat, place with bark side down.
  - Do not use miters—they will move and become unsightly; instead, use an overlapping pattern or, if possible, place all boards parallel to one another.
  - Consider fastening only one side until the wood dries, then fasten the other side.
  - Use quarter-sawn wood if possible.
  - Do not leave a space between boards; a space will widen as the wood shrinks.
- **If wood can be placed dry:**
  - And is cupped, place cupped side down regardless of grain.
  - If wood is flat, place with the best-looking side up.
  - Make sure to leave gaps, especially if wood being placed is below EMC, because it will gain moisture as it approaches EMC and will expand.

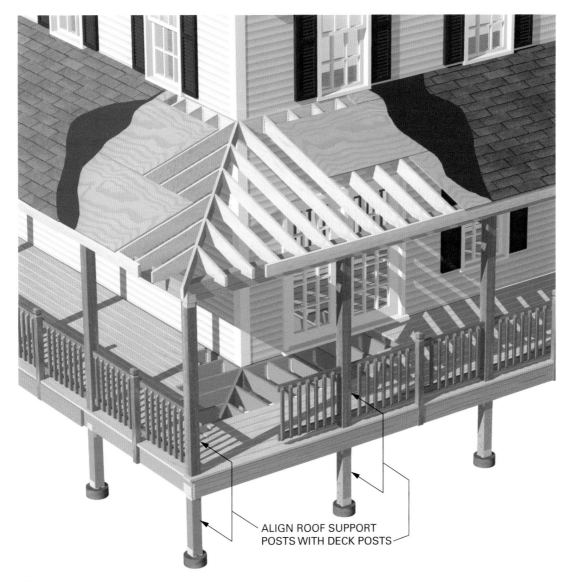

ALIGN ROOF SUPPORT
POSTS WITH DECK POSTS

Figure 7-50 A porch is composed of a deck enclosed by walls and a roof

Covered Porch Construction

When building a covered porch, it is typically constructed on top of the finished deck. Concepts necessary for porch construction, such as post and girder, floor framing, rafter layout, sheathing, etc., have all been covered earlier in the book. At this point following a set of plans should be all that is necessary to continue. Make sure to transfer the roof load downward directly to the deck support posts and pay close attention to moisture and venting issues (this includes roof venting). If hollow posts are used, make sure to allow for air flow all the way through the post. Screening the openings may be necessary to prevent nesting insects from being a problem. Some regions may also require special flashing details to inhibit termites and other insects. See Figure 7-50.

CONSTRUCTION CALCULATORS

Construction calculator use is common today, and even though there are different manufacturers and models, the functions are similar. They are designed to make calculations related to measurements much easier. They have special functions aiding rafter calculations, stairs, board feet, estimating functions, squaring, and much more. Construction calculators will certainly shortcut some number-crunching; however, it is very important that the carpenter fully understands the *concepts* behind these calculations, otherwise the calculations will become nothing more than memorizing a sequence of buttons to push, and the theory will be lost. Theory is very important when trying to figure out a problem that is a little different.

Construction calculator use is recommended. Learning to use one is not difficult and can simplify calculations. Some of the common applications and examples of their ease of use follow. The sequence of buttons being pressed may vary by model and manufacturer.

Construction Calculator Use and Examples

- Conversions—There are four different formats to represent a measurement (feet/inch/fraction, inch/fraction, decimal-feet, and decimal-inches). To enter a measurement, press the number of feet followed by **feet**, the number of inches followed by **inches,** and then the fraction's numerator (top number) followed by /, and last the fraction's denominator. A construction calculator will convert between formats by simply touching **conversion,** followed by either **feet** or **inches**; pressing repeatedly will convert between formats.

- Pythagorean Theorem—The legs of a right triangle are known as the rise and run. Simply enter one leg (measurement) followed by **run**, the other leg followed by **rise,** and then press **diagonal** to obtain the length of the hypotenuse.

- Rafter Calculations—To find the line length of a common rafter, enter the unit rise followed by **pitch**, enter the run followed by **run,** and press **diagonal** to obtain the line length. To find the line length of the corresponding hip or valley rafter, after performing the above operations, simply touch **hip/valley**.

This by no means a comprehensive look at the functions of a construction calculator, and is it **not** intended to be a manual for calculator use. This is simply a brief overview so the reader can realize the value of such a tool. There are several construction calculator apps available for phones.

APPENDIX FOR CHAPTER 2

The following miscellaneous mathematical calculations assume use of a scientific calculator.

Convert Fractions to Decimals

Example 1: Convert ⁵⁄₁₆″ to the decimal equivalent.

Using a calculator, enter 5 ÷ 16 = _____ (answer is .3125″).

Example 2: Convert 3⅞″ to the decimal equivalent.

Using a calculator, enter 7 ÷ 8 = _____ (answer is .875″). **Don't forget** to add the 3″. Final answer is 3.875″.

Convert Decimals to Fractions

Framing carpenters generally work with 8th- or 16th-inch measurements. Therefore, to simplify this operation, multiply the decimal amount by 8 to convert it to eights, or by 16 to convert to sixteenths.

Example 1: Convert .577″ to the nearest 16th. Solution: .577″ × 16 = 9.23 discard the number to the right of the decimal, the answer is ⁹⁄₁₆″.

Example 2: Convert 47.9″ to the nearest 8th. Solution: .9″ × 8 = 7.2 discard number to the right of the decimal and the answer will be 47⅞″.

Convert Feet-Inch Fraction Measurements to the Decimal Equivalent in Feet

Example: Convert 22′ 4¾″ to the decimal equivalent.

- **Step 1**—Convert ¾″ to decimal form, 3 ÷ 4 = .75″. Replace the ¾″ with .75″.
- **Step 2**—Convert 4.75″ to feet, 4.75″ ÷ 12 = .3958′.
- **Step 3**—Add the 22′ back in to give an answer of 22.3958′ = 22′ 4¾″.

Convert Decimal Feet to a Feet-Inch Fraction Measurement

Example: Convert 16.695′ to a feet-inch fraction equivalent.

- **Step 1**—Multiply .695′ by 12 to find the number of inches, .695′ × 12 = 8.34″. Set the 8″ aside.
- **Step 2**—Find out how many sixteenths of an inch .34″ is equivalent to. Multiply .34 × 16 = 5.44—this is closest to ⁵⁄₁₆″.
- **Step 3**—Answer is 16.695′ = 16′ 8⁵⁄₁₆″.

Pythagorean Theorem

Find the length of the diagonal of a rectangular foundation measuring 28′ 6″ × 44′ 9″.

- **Step 1**—Convert the measurements to decimal form:
  28′ 6″ = 6″/12″ = .5′,   28′ + .5′ = **28.5′**
  44′ 9″ = 9/12 = .75′,   44′ + .75′ = **44.75′**

- **Step 2**—Using the Pythagorean Theorem, calculate the length of the diagonal:

$$a^2 + b^2 = c^2$$

$$28.5^2 + 44.75^2 = c^2$$

$$812.25 + 2002.5625 = c^2$$

$$2814.8125 = c^2$$

$$\sqrt{2814.8125} = 53.05480657' = c$$

$$53.0548' = c$$

> *Construction calculators will allow the direct input of fractional measurements.*

Use only the first 3–4 decimal places and round off the last digit.

- **Step 3**—Convert back to inches/16th .0548 × 12 = .6576″ (multiply the fraction by 12 to convert to inches).

Note: There are no whole inches in this example!
Convert to 16th's (multiply the decimal by 16). Solution: .6576 × 16 = 10.52 = ~ 11
Final answer is 53′ ¹¹⁄₁₆″.

APPENDIX FOR CHAPTER 3

Girder and Header Spans

Following is selected information from IRC 2018 Table R602.7(1).

Girder Spans and Header Spans for Exterior Bearing Walls and Required Number of Jack Studs

Maximum spans for Douglas Fir-Larch, Hem-Fir, Southern Pine, and Spruce-Pine-Fir (Minimum #2 grade)

Figure Appendix-1 IRC 2018 Table   R602.7(1)

Girders and Headers Supporting	Size and Quantity	Ground Snow Load (psf)											
		30				50				70			
		Building width in feet											
		24		36		24		36		24		36	
		Span	NJ	Span	NJ	Span	NJ	Span	NJ	Span	NJ	Span	NJ
Roof and Ceiling	1–2 × 6	3–1	2	2–7	2	2–8	2	2–3	2	2–4	2	2–0	2
	1–2 × 8	3–11	2	3–3	2	3–4	2	2–10	2	3–0	2	2–6	3
	1–2 × 10	4–8	2	3–11	2	4–0	2	3–4	2	3–6	3	3–0	3
	1–2 × 12	5–5	2	4–7	3	4–8	3	3–11	3	4–2	3	3–6	3
	2–2 × 6	4–7	1	3–10	1	3–11	1	3–3	2	3–6	2	2–11	2
	2–2 × 8	5–9	1	4–10	2	5–0	2	4–2	2	4–5	2	3–9	2
	2–2 × 10	6–10	2	5–9	2	5–11	2	4–11	2	5–3	2	4–5	2
	2–2 × 12	8–1	2	6–10	2	6–11	2	5–10	2	6–2	2	5–2	3
	3–2 × 8	7–3	1	6–1	1	6–3	1	5–3	2	5–6	2	4–8	2
	3–2 × 10	8–7	1	7–3	2	7–4	2	6–2	2	6–7	2	5–6	2
	3–2 × 12	10–1	2	8–6	2	8–8	2	7–4	2	7–9	2	6–6	2

HEADER, TYP

Figure Appendix-1 *(Continued)*

Girders and Headers Supporting	Size and Quantity	Ground Snow Load (psf)															
		30				50				70						
		Building width in feet														
		24		36		24		36		24		36				
		Span	NJ	Span	NJ	Span	NJ	Span	NJ	Span	NJ	Span	NJ			
Roof, Ceiling, and One Floor (Center–Bearing)	1–2 × 6	2–7	2	2–2	2	2–4	2	2–0	2	2–2	2	1–10	2			
	1–2 × 8	3–3	2	2–9	2	3–0	2	2–6	3	2–9	2	2–4	3			
	1–2 × 10	3–10	2	3–3	3	3–6	3	3–0	3	3–3	3	2–9	3			
	1–2 × 12	4–6	3	3–10	3	4–2	3	3–6	3	3–10	3	3–3	4			
	2–2 × 6	3–9	1	3–3	2	3–6	2	3–0	2	3–3	2	2–9	2			
	2–2 × 8	4–10	2	4–1	2	4–5	2	3–9	2	4–1	2	3–6	2			
	2–2 × 10	5–8	2	4–10	2	5–3	2	4–5	2	4–10	2	4–1	2			
	2–2 × 12	6–8	2	5–8	2	6–2	2	5–3	2	5–8	2	4–10	3			
	3–2 × 8	6–0	1	5–1	2	5–6	2	4–8	2	5–1	2	4–4	2			
	3–2 × 10	7–2	2	6–1	2	6–7	2	5–7	2	6–1	2	5–2	2			
	3–2 × 12	8–5	2	7–2	2	7–8	2	6–7	2	7–1	2	6–1	2			

Spans are given in feet and inches
NJ = number of jacks required to support each end

(continued)

Figure Appendix-1 *(Continued)*

Girders and Headers Supporting	Size and Quantity	Ground Snow Load (psf)											
		30				50				70			
		Building width in feet											
		24		36		24		36		24		36	
		Span	NJ	Span	NJ	Span	NJ	Span	NJ	Span	NJ	Span	NJ
Roof, Ceiling, and One Floor (Clear-Span)	1–2 × 6	2–3	2	1–11	2	2–1	2	1–9	2	2–0	2	1–8	2
	1–2 × 8	2–10	2	2–5	3	2–8	2	2–3	3	2–6	3	2–2	3
	1–2 × 10	3–5	3	2–10	3	3–2	3	2–8	3	3–0	3	2–6	3
	1–2 × 12	4–0	3	3–4	3	3–9	3	3–2	4	3–6	3	3–0	4
	2–2 × 6	3–4	2	2–10	2	3–2	2	2–8	2	3–0	2	2–6	2
	2–2 × 8	4–3	2	3–7	2	4–0	2	3–4	2	3–9	2	3–2	2
	2–2 × 10	5–0	2	4–2	2	4–9	2	4–0	2	4–5	2	3–9	3
	2–2 × 12	5–11	2	4–11	3	5–7	2	4–8	3	5–3	3	4–5	3
	3–2 × 8	5–3	2	4–5	2	5–0	2	4–2	2	4–8	2	4–0	2
	3–2 × 10	6–3	2	5–3	2	5–11	2	5–0	2	5–7	2	4–8	2
	3–2 × 12	7–5	2	6–2	2	7–0	2	5–10	2	6–7	2	5–6	3

Figure Appendix-1 (*Continued*)

Girders and Headers Supporting	Size and Quantity	Ground Snow Load (psf)											
		30				50				70			
		Building width in feet											
		24		36		24		36		24		36	
		Span	NJ	Span	NJ	Span	NJ	Span	NJ	Span	NJ	Span	NJ
Roof, Ceiling, and Two Floors (Center–Bearing)	1–2 × 6	2–1	2	1–10	2	2–0	2	1–9	2	1–11	2	1–8	2
	1–2 × 8	2–9	2	2–4	3	2–7	2	2–2	3	2–5	3	2–1	3
	1–2 × 10	3–2	3	2–9	3	3–1	3	2–7	3	2–11	3	2–5	3
	1–2 × 12	3–9	3	3–2	4	3–7	3	3–1	4	3–5	3	2–11	4
	2–2 × 6	3–2	2	2–8	2	3–0	2	2–7	2	2–10	2	2–5	2
	2–2 × 8	4–0	2	3–5	2	3–10	2	3–3	2	3–7	2	3–1	2
	2–2 × 10	4–9	2	4–0	2	4–6	2	3–10	3	4–3	2	3–8	3
	2–2 × 12	5–7	2	4–9	3	5–4	3	4–6	3	5–0	3	4–3	3
	3–2 × 8	5–0	2	4–3	2	4–9	2	4–1	2	4–6	2	3–10	2
	3–2 × 10	5–11	2	5–1	2	5–8	2	4–10	2	5–4	2	4–7	2
	3–2 × 12	7–0	2	5–11	2	6–8	2	5–8	3	6–4	2	5–4	3

(continued)

Figure Appendix-1 *(Continued)*

Girders and Headers Supporting	Size and Quantity	Ground Snow Load (psf)											
		30				**50**				**70**			
		Building width in feet											
		24		**36**		**24**		**36**		**24**		**36**	
		Span	NJ	Span	NJ	Span	NJ	Span	NJ	Span	NJ	Span	NJ
Roof, Ceiling, and Two Floors (Clear-Span)	1-2 × 6	1-9	2	1-5	2	1-9	2	1-5	3	1-8	2	1-5	3
	1-2 × 8	2-2	3	1-10	3	2-2	3	1-10	3	2-1	3	1-10	3
	1-2 × 10	2-7	3	2-2	3	2-7	3	2-2	4	2-6	3	2-2	4
	1-2 × 12	3-0	3	2-7	4	3-0	4	2-7	4	3-0	4	2-6	4
	2-2 × 6	2-6	2	2-2	2	2-6	2	2-2	2	2-6	2	2-1	2
	2-2 × 8	3-3	2	2-8	2	3-3	2	2-8	2	3-2	2	2-8	3
	2-2 × 10	3-10	2	3-2	3	3-10	3	3-2	3	3-9	3	3-2	3
	2-2 × 12	4-6	3	3-9	3	4-6	3	3-9	3	4-5	3	3-9	3
	3-2 × 8	4-0	2	3-5	2	4-0	2	3-5	2	3-11	2	3-4	2
	3-2 × 10	4-9	2	4-0	2	4-9	2	4-0	2	4-8	2	4-0	3
	3-2 × 12	5-8	2	4-9	3	5-8	3	4-9	3	5-6	3	4-8	3

APPENDIX FOR CHAPTER 4

Floor Joist Spans

Following is selected information from IRC 2018 Table R502.3.1(1).

Figure Appendix-2 IRC 2018 Table R502.3.1(1)

Floor Joist Spans for Common Lumber Species
Residential **sleeping** areas, live load = **30** psf, L/Δ = 360

12" OC Framing		DEAD LOAD = 10 psf				DEAD LOAD = 20 psf			
Species	Grade	2 × 6	2 × 8	2 × 10	2 × 12	2 × 6	2 × 8	2 × 10	2 × 12
Southern Pine	#1	11–10	15–7	19–10	24–2	11–10	15–7	18–7	22–0
	#2	11–3	14–11	18–1	21–4	10–9	13–8	16–2	19–1
	#3	11–3	11–6	14–0	16–6	8–2	10–3	12–6	14–9
Spruce/Pine/Fir	#1	11–3	14–11	19–0	23–0	11–3	14–7	17–9	20–7
	#2	11–3	14–11	19–0	23–0	11–3	14–7	17–9	20–7
	#3	9–8	12–4	15–0	17–5	8–8	11–0	13–5	15–7
Douglas Fir-Larch	#1	12–0	15–10	20–3	24–8	12–0	15–7	19–0	22–0
	#2	11–0	15–7	19–10	23–4	11–8	14–9	18–0	20–11
	#3	9–11	12–7	15–5	17–10	8–11	11–3	13–9	16–0
Hem-Fir	#1	11–7	15–3	19–5	23–7	11–7	15–3	18–9	21–9
	#2	11–0	14–6	18–6	22–6	11–0	14–4	17–6	20–4
	#3	9–8	12–4	15–0	17–5	8–8	11–0	13–5	15–7

Floor Joist Spans for Common Lumber Species
Residential **sleeping** areas, live load = **30** psf, L/Δ = 360

16" OC Framing		DEAD LOAD = 10 psf				DEAD LOAD = 20 psf			
Species	Grade	2 × 6	2 × 8	2 × 10	2 × 12	2 × 6	2 × 8	2 × 10	2 × 12
Southern Pine	#1	10–9	14–2	18–0	21–4	10–9	13–9	16–1	19–1
	#2	10–3	13–3	15–8	18–6	9–4	11–10	14–0	16–6
	#3	7–11	10–0	11–1	14–4	7–1	8–11	10–10	12–10
Spruce/Pine/Fir	#1	10–3	13–6	17–2	19–11	9–11	12–7	15–5	17–10
	#2	10–3	13–6	17–2	19–11	9–11	12–7	15–5	17–10
	#3	8–5	10–8	13–0	15–1	7–6	9–6	11–8	13–6
Douglas Fir-Larch	#1	10–11	14–5	18–5	21–4	10–8	13–6	16–5	19–1
	#2	10–9	14–2	17–5	20–3	10–1	12–9	15–7	18–1
	#3	8–7	10–11	13–4	15–5	7–8	9–9	11–11	13–10
Hem-Fir	#1	10–6	13–10	17–8	21–1	10–6	13–4	16–3	18–10
	#2	10–0	13–2	16–10	19–8	9–10	12–5	15–2	17–7
	#3	8–5	10–8	13–0	15–1	7–6	9–6	11–8	13–6

(continued)

Figure Appendix-2 (*Continued*)

Floor Joist Spans for Common Lumber Species
Residential **sleeping** areas, live load = **30** psf, L/Δ = 360

24" OC Framing		DEAD LOAD = 10 psf				DEAD LOAD = 20 psf			
Species	Grade	2 × 6	2 × 8	2 × 10	2 × 12	2 × 6	2 × 8	2 × 10	2 × 12
Southern Pine	#1	9–4	12–4	14–8	17–5	8–10	11–3	13–1	15–7
	#2	8–6	10–10	12–10	15–1	7–7	9–8	11–5	13–6
	#3	6–5	8–2	9–10	11–8	5–9	7–3	8–10	10–5
Spruce/Pine/Fir	#1	8–11	11–6	14–1	16–3	8–1	10–3	12–7	14–7
	#2	8–11	11–6	14–1	16–3	8–1	10–3	12–7	14–7
	#3	6–10	8–8	10–7	12–4	6–2	7–9	9–6	11–0
Douglas Fir-Larch	#1	9–7	12–4	15–0	17–5	8–8	11–0	13–5	15–7
	#2	9–3	11–8	14–3	16–6	8–3	10–5	12–9	14–9
	#3	7–0	8–11	10–11	12–7	6–3	8–0	9–9	11–3
Hem-Fir	#1	9–2	12–1	14–10	17–2	8–7	10–10	13–3	15–5
	#2	8–9	11–4	13–10	16–1	8–0	10–2	12–5	14–4
	#3	6–10	8–8	10–7	12–4	6–2	7–9	9–6	11–0

Floor Joist Spans for Common Lumber Species
Residential **living** areas, live load = **40** psf, L/Δ = 360

12" OC Framing		DEAD LOAD = 10 psf				DEAD LOAD = 20 psf			
Species	Grade	2 × 6	2 × 8	2 × 10	2 × 12	2 × 6	2 × 8	2 × 10	2 × 12
Southern Pine	#1	10–9	14–2	18–0	21–11	10–9	14–2	16–11	20–1
	#2	10–3	13–6	16–2	19–1	9–10	12–6	14–9	17–5
	#3	8–2	10–3	12–6	14–9	7–5	9–5	11–5	13–6
Spruce/Pine/Fir	#1	10–3	13–6	17–3	20–7	10–3	13–3	16–3	18–10
	#2	10–3	13–6	17–3	20–7	10–3	13–3	16–3	18–10
	#3	8–8	11–0	13–5	15–7	7–11	10–0	12–3	14–3
Douglas Fir-Larch	#1	10–11	14–5	18–5	22–0	10–11	14–2	17–4	20–1
	#2	10–9	14–2	18–0	20–11	10–8	13–6	16–5	19–1
	#3	8–11	11–3	13–9	16–0	8–1	10–3	12–7	14–7
Hem-Fir	#1	10–6	13–10	17–8	21–6	10–6	13–10	17–1	19–10
	#2	10–0	13–2	16–10	20–4	10–0	13–1	16–0	18–6
	#3	8–8	11–0	13–5	15–7	7–11	10–0	12–3	14–3

Figure Appendix-2 (*Continued*)

Floor Joist Spans for Common Lumber Species
Residential **living** areas, live load = **40** psf, L/Δ = 360

16" OC Framing		DEAD LOAD = 10 psf				DEAD LOAD = 20 psf			
Species	Grade	2 × 6	2 × 8	2 × 10	2 × 12	2 × 6	2 × 8	2 × 10	2 × 12
Southern Pine	#1	9–9	12–10	16–1	19–1	9–9	12–7	14–8	17–5
	#2	9–4	11–10	14–0	16–6	8–6	10–10	12–10	15–1
	#3	7–1	8–11	10–10	12–10	6–5	8–2	9–10	11–8
Spruce/Pine/Fir	#1	9–4	12–3	15–5	17–10	9–1	11–6	14–1	16–3
	#2	9–4	12–3	15–5	17–10	9–1	11–6	14–1	16–3
	#3	7–6	9–6	11–8	13–6	6–10	8–8	10–7	12–4
Douglas Fir-Larch	#1	9–11	13–1	16–5	19–1	9–8	12–4	15–0	17–5
	#2	9–9	12–9	15–7	18–1	9–3	11–8	14–3	16–6
	#3	7–8	9–9	11–11	13–10	7–0	8–11	10–11	12–7
Hem-Fir	#1	9–6	12–7	16–0	18–10	9–6	12–2	14–10	17–2
	#2	9–1	12–0	15–2	17–7	8–11	11–4	13–10	16–1
	#3	7–6	9–6	11–8	13–6	6–10	8–8	10–7	12–4

Floor Joist Spans for Common Lumber Species
Residential **living** areas, live load = **40** psf, L/Δ = 360

24" OC Framing		DEAD LOAD = 10 psf				DEAD LOAD = 20 psf			
Species	Grade	2 × 6	2 × 8	2 × 10	2 × 12	2 × 6	2 × 8	2 × 10	2 × 12
Southern Pine	#1	8–6	11–3	13–1	15–7	8–1	10–3	12–0	14–3
	#2	7–7	9–8	11–5	13–6	7–0	8–10	10–5	12–4
	#3	5–9	7–3	8–10	10–5	5–3	6–8	8–1	9–6
Spruce/Pine/Fir	#1	8–1	10–3	12–7	14–7	7–5	9–5	11–6	13–4
	#2	8–1	10–3	12–7	14–7	7–5	9–5	11–6	13–4
	#3	6–2	7–9	9–6	11–0	5–7	7–1	8–8	10–1
Douglas Fir-Larch	#1	8–8	11–0	13–5	15–7	7–11	10–0	12–3	14–3
	#2	8–3	10–5	12–9	14–9	7–6	9–6	11–8	13–6
	#3	6–3	8–0	9–9	11–3	5–9	7–3	8–11	10–4
Hem-Fir	#1	8–4	10–10	13–3	15–5	7–10	9–11	12–1	14–0
	#2	7–11	10–2	12–5	14–4	7–4	9–3	11–4	13–1
	#3	6–2	7–9	9–6	11–0	5–7	7–1	8–8	10–1

APPENDIX FOR CHAPTER 5

Fastening Schedule

Following is selected information from IRC 2018 Table R602.3(1).

For more complete information, including other types of fasteners such as box or deformed nails, see IRC 2018 International Codes.

Figure Appendix-3 IRC 2018 Table R602.3(1)

Item	Description of Building Elements	Number and Type of Fastener	Spacing and Location
Roof			
1	Blocking between ceiling joists or rafters to top plate	3-8d common (2½" × 0.131"); or 3-3" × 0.131" nails	Toe nail
2	Ceiling joists to top plate	3-8d common (2½" × 0.131"); or 3-3" × 0.131" nails	Per joist, toe nail
3	Ceiling joist not attached to parallel rafter, laps over partitions	3-16d common (3½" × 0.162); or 4-3" × 0.131" nails	Face nail
4	Ceiling joist attached to parallel rafter (heel joint)	Refer to table R802.5.2 in IRC 2018	Face nail
5	Collar tie to rafter, face nail or 1¼" x 20 gage. Ridge strap to rafter	3-10d common (3" × 0.148") or; 4-3" × 0.131" nails	Face nail each rafter
6	Rafter of roof truss to plate	3-10d common (3" × 0.148") or; 4-3" × 0.131" nails	2 toe nails on one side and 1 toe nail on opposite side of each rafter or truss
7	Roof rafters to ridge, valley or hip rafters or roof rafter to minimum 2" ridge beam	2-16d common (3½" × 0.162); or 3-3" × 0.131" nails	End nail
Wall			
8	Stud to stud (not at braced wall panels)	16d common (3½" × 0.162)	24" OC face nail
		3" x 0.131" nails	16" OC face nail
9	Stud to stud and abutting studs at intersecting wall corners (at braced wall panels)	16d common (3½" × 0.162); or 3" × 0.131" nails	12" OC face nail
		16d common (3½" × 0.162)	16" OC face nail
10	Built-up header (2" to 2" header with ½" spacer)	16d common (3½" × 0.162)	16" OC each edge face nail

Figure Appendix-3 (Continued)

Item	Description of Building Elements	Number and Type of Fastener	Spacing and Location
Wall			
11	Continuous header to stud	4-8d common (2½" × 0.131")	Toe nail
12	Top plate to top plate	16d common (3½" × 0.162); or	16" OC face nail
		3" × 0.131" nails	12" OC face nail
13	Double top plate splice	8-16d common (3½" × 0.162); or 12-3" c 0.131" nails	Face nail on each side of end joint (minimum 24" lap splice length each side of end joint)
14	Bottom plate to joist, rim joist, band joist or blocking (not at braced wall panels)	16d common (3½" × 0.162)	16" OC face nail
		3" × 0.131" nails	12" OC face nails
15	Bottom plate to joist, rim joist, band joist or blocking (at braced wall panels)	2-16d common (3½" × 0.162); or	2 each 16" OC face nail
		4-3" × 0.131" nails	4 each 16" OC face nail
16	Top or bottom plate to stud	4-8d common (2½" × 0.131); or 4-3" × 0.131" nails	Toe nail
		2-16d common (3½" × 0.162); or 3-3" × 0.131" nails	End nail
17	Top plates, laps at corners, and intersections	2-16d common (3½" × 0.162); or 3-3" × 0.131" nails	Face nail
18	1" brace to each stud and plate	2-8d common (2½" × 0.113"); or 2 staples 1¾"	Face nail
19	1" x 6" sheathing to each bearing	2-8d common (2½" × 0.113"); or 2 staples, 1" crown, 16 gage, 1¾"	Face nail
20	1" x 8" and wider sheathing to each bearing	3-8d common (2½" × 0.113"); or 3 staples, 1" crown, 16 gage, 1¾"	Face nail
		Wider than 8" 3-8d common (2½" × 0.113"); or 4 staples, 1" crown, 16 gage, 1¾"	
Floor			
21	Joist to sill, top plate, or girder	3-8d common (2½" x 0.113"); or 3-3" x 0.131" nails	Toe nail

(continued)

Figure Appendix-3 (*Continued*)

Item	Description of Building Elements	Number and Type of Fastener	Spacing and Location	
22	Rim joist, band joist or blocking to sill, or top plate (roof applications also)	8d common (2½" × 0.113"); or 3-3" × 0.131" nails	6" OC toe nail	
23	1" × 6" subfloor or less to each joist	2-8d common (2½" × 0.113"); or 2 staples, 1" crown, 16 gage, 1¾"	Face nail	
24	2" subfloor to joist or girder	2-16d common (3½" × 0.162")	Blind and face nail	
25	2" plank (plank & beam-floor & roof)	2-16d common (3½" × 0.162")	At each bearing, face nail	
26	Band or rim joist to joist	3-16d common (3½" × 0.162"); or 4-3" × 0.131" nails; or 4-3" × 14 gage, staples 7/16" crown	End nail	
27	Built-up girders and beams, 2" lumber layers	20d common (4" × 0.192"); or	Nail each layer 32" OC at top and bottom and staggered	
		3" × 0.131" nails; or	24" OC face nail at top and bottom staggered on opposite sides	
		2-20d common (4" × 0.192"); or 3-3" × 0.131" nails	Face nail at ends and at each splice	
28	Ledger strip supporting joists or rafters	3-16d common (3½" × 0.162"); or 4-3" × 0.131" nails	At each joist or rafter, face nail	
29	Bridging of blocking to joist	2-8d common (2½" × 0.131"); or 2-3" × 0.131" nails	Each end, toe nail	

Item	Description of Building Element	Number and Type of Fastener	Spacing of Fasteners	
			Edges (in)	Intermediate Supports (in)
Wood structural panels, subfloor, roof, and interior wall sheathing to framing and particleboard wall sheathing to framing (See IRC 2018 Table R602.3(3) for more information on exterior sheathing to wall framing requirements.)				
30	3/8"–1/2"	6d common (2" × 0.113") nail (subfloor to wall) 8d common (2½" × 0.131") nail (roof)	6	12
31	19/32"–1"	8d common (2½" × 0.131") nail (roof)	6	12

Figure Appendix-3 (*Continued*)

32	$1\frac{1}{8}$"–$1\frac{1}{4}$"	10d common ($2\frac{1}{2}$" × 0.148")	6	12

Other wall sheathing

33	$\frac{1}{2}$" structural cellulosic fiberboard sheathing	$1\frac{1}{2}$" galvanized roofing nail, $^7/_{16}$" head diameter, or $1\frac{1}{4}$" long 16 gage staple with $^7/_{16}$" or 1" crown	3	6
34	$^{25}/_{32}$" structural cellulosic fiberboard sheathing	$1\frac{3}{4}$" galvanized roofing nail, $^7/_{16}$" head diameter, or $1\frac{1}{2}$" long 16 gage Staple with $^7/_{156}$" or 1" crown	3	6
35	$\frac{1}{2}$" gypsum sheathing	$1\frac{1}{2}$" galvanized roofing nail; staple galvanized, $1\frac{1}{2}$" long; $1\frac{1}{4}$" screws, Type W or S	7	7
36	$^5/_8$" gypsum sheathing	$1\frac{3}{4}$" galvanized roofing nail; staple galvanized, $1^5/_8$" long screws, Type W or S	7	7

Wood structural panels, combination subfloor underlayment to framing

37	$\frac{3}{4}$" and less	8d common ($2\frac{1}{2}$" × 0.131") nail	6	12
38	$^7/_8$"–1"	8d common ($2\frac{1}{2}$" × 0.131") nail	6	12
39	$1\frac{1}{8}$"–$1\frac{1}{4}$"	10d common (3" × 0.148") nail	6	12

Table excerpted from the 2018 International Residential Code; Copyright 2017. Washington, D.C.: International Code Council. Reproduced with permission. All rights reserved. www.ICCSAFE.org

Figure Appendix-4 Following is selected information from IRC 2018 Table 602.3(3).
Requirements for Wood Structural Panel Wall Sheathing Used to Resist Wind Pressures

Minimum Nail		Minimum Wood Structural Panel Span Rating (in)	Minimum Nominal Panel Thickness (in)	Maximum Wall Stud Spacing (in)	Panel Nail Spacing (in)		Ultimate Design Wind Speed (mph)		
Size	Penetration (in)				Edges (in oc)	Field (in oc)	Wind Exposure Category		
							B	C	D
6d common (2" × 0.113")	1.5	24/0	$^3/_8$	16	6	12	140	115	110
8d common (2.5" × 0.131")	1.75	24/16	$^7/_{16}$	16	6	12	170	140	135
				24	6	12	140	115	110

APPENDIX FOR CHAPTER 6

Rafter spans

Following is selected from IRC 2018 Tables R802.4.1(2)–R802.1.4(8).

Assumes ceiling joists are located at the bottom of the attic space or that some other method of resisting the outward push of the rafters on the bearing walls is provided at that location.

Figure Appendix-5 IRC 2018 Table   *R802.4.1(2)*

Rafter Spans for Common Lumber Species Roof live load = **20** psf, ceiling attached to rafters, L/Δ = 240									
12″ OC Framing		DEAD LOAD = 10 psf				DEAD LOAD = 20 psf			
Species	Grade	2 × 6	2 × 8	2 × 10	2 × 12	2 × 6	2 × 8	2 × 10	2 × 12
Douglas Fir-Larch	#1	15–9	20–10	26+	26+	15–4	19–5	23–9	26+
	#2	15–6	20–5	26–0	26+	14–7	18–5	22–6	26–0
	#3	12–10	16–3	19–10	23–0	11–1	14–1	17–2	19–11
Hem-Fir	#1	15–2	19–11	25–5	26+	15–2	19–2	23–5	26+
	#2	14–5	19–0	24–3	26+	14–2	17–11	21–11	25–5
	#3	12–6	15–10	19–5	22–6	10–10	13–9	16–9	19–6
Southern Pine	#1	15–6	20–5	26+	26+	15–6	19–10	23–2	26+
	#2	14–9	19–6	23–5	26+	13–6	17–1	20–3	23–10
	#3	11–9	14–10	18–0	21–4	10–2	12–10	15–7	18–6
Spruce-Pine-Fir	#1	14–9	19–6	24–10	26+	14–4	18–2	22–3	25–9
	#2	14–9	19–6	24–10	26+	14–4	18–2	22–3	25–9
	#3	12–6	15–10	19–5	22–6	10–10	13–9	16–9	19–6

Rafter Spans for Common Lumber Species
Roof live load = **20** psf, ceiling attached to rafters, L/Δ = 240

16" OC Framing		DEAD LOAD = 10 psf				DEAD LOAD = 20 psf			
Species	Grade	2 × 6	2 × 8	2 × 10	2 × 12	2 × 6	2 × 8	2 × 10	2 × 12
Douglas Fir-Larch	#1	14–4	18–11	23–9	26+	13–3	16–10	20–7	23–10
	#2	14–1	18–5	22–6	26–0	12–7	16–0	19–6	22–7
	#3	11–7	14–1	17–2	19–11	9–8	12–2	14–11	17–3
Hem-Fir	#1	13–9	18–1	23–1	26+	13–1	16–7	20–4	23–7
	#2	13–1	17–3	21–11	25–5	12–3	15–6	18–11	22–0
	#3	10–10	13–9	16–9	19–6	9–5	11–11	14–6	16–10
Southern Pine	#1	14–1	18–6	23–2	26+	13–7	17–2	20–1	23–10
	#2	13–5	17–1	20–3	23–10	11–8	14–9	17–6	20–8
	#3	10–2	12–10	15–7	18–6	8–10	11–2	13–6	16–0
Spruce-Pine-Fir	#1	13–5	17–9	22–3	25–9	12–5	15–9	19–3	22–4
	#2	13–5	17–9	22–3	25–9	12–5	15–9	19–3	22–4
	#3	10–10	13–9	16–9	19–6	9–5	11–11	14–6	16–10

Rafter Spans for Common Lumber
Roof live load = **20** psf, ceiling attached to rafters, L/Δ = 240

24" OC Framing		DEAD LOAD = 10 psf				DEAD LOAD = 20 psf			
Species	Grade	2 × 6	2 × 8	2 × 10	2 × 12	2 × 6	2 × 8	2 × 10	2 × 12
Douglas Fir-Larch	#1	12–6	15–10	19–5	22–6	10–10	13–9	16–9	19–6
	#2	11–11	15–1	18–5	21–4	10–4	13–0	15–11	18–6
	#3	9–1	11–6	14–1	16–3	7–10	10–0	12–2	14–1
Hem-Fir	#1	12–0	15–8	19–2	22–2	10–9	13–7	16–7	19–3
	#2	11–5	14–8	17–10	20–9	10–0	12–8	15–6	17–11
	#3	8–10	11–3	13–8	15–11	7–8	9–9	11–10	13–9
Southern Pine	#1	12–3	16–2	18–11	22–6	11–1	14–0	16–5	19–6
	#2	11–0	13–11	16–6	19–6	9–6	12–1	14–4	16–10
	#3	8–4	10–6	12–9	15–1	7–3	9–1	11–0	13–1
Spruce-Pine-Fir	#1	11–9	14–10	18–2	21–0	10–2	12–10	15–8	18–3
	#2	11–9	14–10	18–2	21–0	10–2	12–10	15–8	18–3
	#3	8–10	11–3	13–8	15–11	7–8	9–9	11–10	13–9

Figure Appendix-6 IRC 2018 Table   R802.4.1 2(4)

Rafter Spans for Common Lumber
Roof live load = **30** psf, ceiling attached to rafters, L/Δ = 240

12″ OC Framing		DEAD LOAD = 10 psf				DEAD LOAD = 20 psf			
Species	Grade	2 × 6	2 × 8	2 × 10	2 × 12	2 × 6	2 × 8	2 × 10	2 × 12
Douglas Fir-Larch	#1	13–9	18–2	22–9	26+	13–2	16–8	20–4	23–7
	#2	13–6	17–8	21–7	25–1	12–6	15–10	19–4	22–5
	#3	10–8	13–6	16–6	19–2	9–6	12–1	14–9	17–1
Hem-Fir	#1	13–3	17–5	22–3	26–0	13–0	16–6	20–1	23–4
	#2	12–7	16–7	21–0	24–4	12–2	15–4	18–9	21–9
	#3	10–5	13–2	16–1	18–8	9–4	11–9	14–5	16–8
Southern Pine	#1	13–6	17–10	22–3	26+	13–5	17–0	19–11	23–7
	#2	12–11	16–4	19–5	22–10	11–7	14–8	17–4	20–5
	#3	9–9	12–4	15–0	17–9	8–9	11–0	13–5	15–10
Spruce-Pine-Fir	#1	12–11	17–0	21–4	24–8	21–4	15–7	19–1	22–1
	#2	12–11	17–0	21–4	24–8	21–4	15–7	19–1	22–1
	#3	10–5	13–2	16–1	18–8	9–4	11–9	14–5	16–8

Rafter Spans for Common Lumber
Roof live load = **30** psf, ceiling attached to rafters, L/Δ = 240

16″ OC Framing		DEAD LOAD = 10 psf				DEAD LOAD = 20 psf			
Species	Grade	2 × 6	2 × 8	2 × 10	2 × 12	2 × 6	2 × 8	2 × 10	2 × 12
Douglas Fir-Larch	#1	12–6	16–2	19–9	22–10	11–5	14–5	17–8	20–5
	#2	12–1	15–4	18–9	21–8	10–10	13–8	16–9	19–5
	#3	9–3	11–8	14–3	16–7	8–3	10–6	12–9	14–10
Hem-Fir	#1	12–0	15–10	19–6	22–7	11–3	14–3	17–5	20–2
	#2	11–5	14–11	18–2	21–11	10–6	13–4	16–3	18–10
	#3	9–0	11–5	13–11	16–2	8–1	10–3	12–6	14–6
Southern Pine	#1	12–3	16–2	19–3	22–10	11–7	14–9	17–3	20–5
	#2	11–2	14–2	16–10	19–10	10–0	12–8	15–1	17–9
	#3	8–6	10–8	13–0	15–4	7–7	9–7	11–7	13–9
Spruce-Pine-Fir	#1	11–9	15–1	18–5	21–5	10–8	13–6	16–6	19–2
	#2	11–9	15–1	18–5	21–5	10–8	13–6	16–6	19–2
	#3	9–0	11–5	13–11	16–2	8–1	10–3	12–6	14–6

Rafter Spans for Common Lumber
Roof live load = **30** psf, ceiling attached to rafters, L/Δ = 240

24" OC Framing		DEAD LOAD = 10 psf				DEAD LOAD = 20 psf			
Species	Grade	2 × 6	2 × 8	2 × 10	2 × 12	2 × 6	2 × 8	2 × 10	2 × 12
Douglas Fir-Larch	#1	10–5	13–2	16–1	18–8	9–4	11–9	14–5	16–8
	#2	9–10	12–6	15–3	17–9	8–10	11–2	13–8	15–10
	#3	7–7	9–7	11–8	13–6	6–9	8–7	10–5	12–1
Hem-Fir	#1	10–3	13–0	15–11	18–5	9–2	11–8	14–3	16–6
	#2	9–7	12–2	14–10	17–3	8–7	10–10	13–3	15–5
	#3	7–4	9–4	11–5	13–2	6–7	8–4	10–2	11–10
Southern Pine	#1	10–7	13–5	15–9	18–8	9–6	12–0	14–1	16–8
	#2	9–2	11–7	13–9	16–2	8–2	10–4	12–3	14–6
	#3	6–11	8–9	10–7	12–6	6–2	7–10	9–6	11–2
Spruce-Pine-Fir	#1	9–9	12–4	15–1	17–6	8–8	11–0	13–6	15–7
	#2	9–9	12–4	15–1	17–6	8–8	11–0	13–6	15–7
	#3	7–4	9–4	11–5	13–2	6–7	8–4	10–2	11–10

Table excerpted from the 2018 International Residential Code; Copyright 2017. Washington, D.C.: International Code Council. Reproduced with permission. All rights reserved. www.ICCSAFE.org

Figure Appendix-7 IRC 2018 Table   R802.4.1(6)

Rafter Spans for Common Lumber
Roof live load = **50** psf, ceiling attached to rafters, L/Δ = 240

12" OC Framing		DEAD LOAD = 10 psf				DEAD LOAD = 20 psf			
Species	Grade	2 × 6	2 × 8	2 × 10	2 × 12	2 × 6	2 × 8	2 × 10	2 × 12
Douglas Fir-Larch	#1	11–7	15–3	18–7	21–7	11–2	14–1	17–3	20–0
	#2	11–5	14–5	17–8	20–5	10–7	13–4	16–4	18–11
	#3	8–9	11–0	13–6	15–7	8–1	10–3	12–6	14–6
Hem-Fir	#1	11–2	14–8	18–4	21–3	11–0	13–11	17–0	19–9
	#2	10–8	14–0	17–2	19–11	10–3	13–0	15–10	18–5
	#3	8–6	10–9	13–2	15–3	7–10	10–0	12–2	14–1
Southern Pine	#1	11–5	15–0	18–2	21–7	11–4	14–5	16–10	20–0
	#2	10–6	13–4	15–10	18–8	9–9	12–4	14–8	17–3
	#3	8–0	10–1	12–3	14–6	7–5	9–4	11–4	13–5
Spruce-Pine-Fir	#1	10–11	14–3	17–5	20–2	10–5	13–2	16–1	18–8
	#2	10–11	14–3	17–5	20–2	10–5	13–2	16–1	18–8
	#3	8–6	10–9	13–2	15–3	7–10	10–0	12–2	14–1

Rafter Spans for Common Lumber
Roof live load = **50** psf, ceiling attached to rafters, L/Δ = 240

16" OC Framing		DEAD LOAD = 10 psf				DEAD LOAD = 20 psf			
Species	Grade	2 × 6	2 × 8	2 × 10	2 × 12	2 × 6	2 × 8	2 × 10	2 × 12
Douglas Fir-Larch	#1	10–5	13–2	16–1	18–8	9–8	12–2	14–11	17–3
	#2	9–10	12–6	15–3	17–9	9–2	11–7	14–2	16–5
	#3	7–7	9–7	11–8	13–6	7–0	8–10	10–10	12–6
Hem-Fir	#1	10–2	13–0	15–11	18–5	9–6	12–1	14–9	17–1
	#2	9–7	12–2	14–10	17–3	8–11	11–3	13–9	15–11
	#3	7–4	9–4	11–5	13–2	6–10	8–8	10–6	12–3
Southern Pine	#1	10–4	13–5	15–9	18–8	9–10	12–5	14–7	17–3
	#2	9–2	11–7	13–9	16–2	8–5	10–9	12–9	15–0
	#3	6–11	8–9	10–7	12–6	6–5	8–1	9–10	11–7
Spruce-Pine-Fir	#1	9–9	12–4	15–1	17–6	9–0	11–5	13–11	16–2
	#2	9–9	12–4	15–1	17–6	9–0	11–5	13–11	16–2
	#3	7–4	9–4	11–5	13–2	6–10	8–8	10–6	12–3

Rafter Spans for Common Lumber
Roof live load = **50** psf, ceiling attached to rafters, L/Δ = 240

24" OC Framing		DEAD LOAD = 10 psf				DEAD LOAD = 20 psf			
Species	Grade	2 × 6	2 × 8	2 × 10	2 × 12	2 × 6	2 × 8	2 × 10	2 × 12
Douglas Fir-Larch	#1	8–6	10–9	13–2	15–3	7–10	10–0	12–2	14–1
	#2	8–1	10–3	12–6	14–6	7–6	9–5	11–7	13–5
	#3	6–2	7–10	9–6	11–1	5–8	7–3	8–10	10–3
Hem-Fir	#1	8–5	10–8	13–0	15–1	7–9	9–10	12–0	13–11
	#2	7–10	9–11	12–1	14–1	7–3	9–2	11–3	13–0
	#3	6–0	7–7	9–4	10–9	5–7	7–1	8–7	10–0
Southern Pine	#1	8–8	11–0	12–10	15–3	8–0	10–2	11–11	14–1
	#2	7–5	9–5	11–3	13–2	6–11	8–9	10–5	12–3
	#3	5–8	7–1	8–8	10–3	5–3	6–7	8–0	9–6
Spruce-Pine-Fir	#1	7–11	10–1	12–4	14–3	7–4	9–4	11–5	13–2
	#2	7–11	10–1	12–4	14–3	7–4	9–4	11–5	13–2
	#3	6–0	7–7	9–4	10–9	5–7	7–1	8–7	10–0

Figure Appendix-8 R802.4.1(8)

Rafter Spans for Common Lumber
Roof live load = **70** psf, ceiling attached to rafters, L/Δ = 240

12″ OC Framing		DEAD LOAD = 10 psf				DEAD LOAD = 20 psf			
Species	Grade	2 × 6	2 × 8	2 × 10	2 × 12	2 × 6	2 × 8	2 × 10	2 × 12
Douglas Fir-Larch	#1	10–5	13–2	16–1	18–8	9–10	12–5	15–2	17–7
	#2	9–10	12–6	15–3	17–9	9–4	11–9	14–5	16–8
	#3	7–7	9–7	11–8	13–6	7–1	9–0	11–0	12–9
Hem-Fir	#1	10–0	13–0	15–11	18–5	9–8	12–3	15–0	17–5
	#2	9–6	12–2	14–10	17–3	9–1	11–5	14–0	16–3
	#3	7–4	9–4	11–5	13–2	6–11	8–9	10–9	12–5
Southern Pine	#1	10–2	13–5	15–9	18–8	10–0	12–8	14–10	17–7
	#2	9–2	11–7	13–9	16–2	8–7	10–11	12–11	15–3
	#3	6–11	8–9	10–7	12–6	6–6	8–3	10–0	11–10
Spruce-Pine-Fir	#1	9–9	12–4	15–1	17–6	9–2	11–8	14–2	16–6
	#2	9–9	12–4	15–1	17–6	9–2	11–8	14–2	16–6
	#3	7–4	9–4	11–5	13–2	6–11	8–9	10–9	12–5

Rafter Spans for Common Lumber
Roof live load = **70** psf, ceiling attached to rafters, L/Δ = 240

16″ OC Framing		DEAD LOAD = 10 psf				DEAD LOAD = 20 psf			
Species	Grade	2 × 6	2 × 8	2 × 10	2 × 12	2 × 6	2 × 8	2 × 10	2 × 12
Douglas Fir-Larch	#1	9–0	11–5	13–11	16–2	8–6	10–9	13–2	15–3
	#2	8–7	10–10	13–3	15–4	8–1	10–3	12–6	14–6
	#3	6–6	8–3	10–1	11–9	6–2	7–10	9–6	11–1
Hem-Fir	#1	8–11	11–3	13–9	16–0	8–5	10–8	13–0	15–1
	#2	8–4	10–6	12–10	14–11	7–10	9–11	12–1	14–1
	#3	6–4	8–1	9–10	11–5	6–0	7–7	9–4	10–9
Southern Pine	#1	9–2	11–8	13–8	16–2	8–8	11–0	12–10	15–3
	#2	7–11	10–0	11–11	14–0	7–5	9–5	11–3	13–2
	#3	6–0	7–7	9–2	10–10	5–8	7–1	8–8	10–3
Spruce-Pine-Fir	#1	8–5	10–8	13–1	15–2	7–11	10–1	12–4	14–3
	#2	8–5	10–8	13–1	15–2	7–11	10–1	12–4	14–3
	#3	6–4	8–1	9–10	11–5	6–0	7–7	9–4	10–9

Rafter Spans for Common Lumber
Roof live load = **70** psf, ceiling attached to rafters, L/Δ = 240

24" OC Framing		DEAD LOAD = 10 psf				DEAD LOAD = 20 psf			
Species	Grade	2 × 6	2 × 8	2 × 10	2 × 12	2 × 6	2 × 8	2 × 10	2 × 12
Douglas Fir-Larch	#1	7–4	9–4	11–5	13–2	6–11	8–9	10–9	12–5
	#2	7–0	8–10	10–10	12–6	6–7	8–4	10–2	11–10
	#3	5–4	6–9	8–3	9–7	5–0	6–4	7–9	9–0
Hem-Fir	#1	7–3	9–2	11–3	13–0	6–10	8–8	10–7	12–4
	#2	6–9	8–7	10–6	12–2	6–5	8–1	9–11	11–6
	#3	5–2	6–7	8–1	9–4	4–11	6–3	7–7	8–10
Southern Pine	#1	7–6	9–6	11–1	13–2	7–1	9–0	10–6	12–5
	#2	6–5	8–2	9–9	11–5	6–1	7–9	9–2	10–9
	#3	4–11	6–2	7–6	8–10	4–7	5–10	7–1	8–4
Spruce-Pine-Fir	#1	6–11	8–9	10–8	12–4	6–6	8–3	10–0	11–8
	#2	6–11	8–9	10–8	12–4	6–6	8–3	10–0	11–8
	#3	5–2	6–7	8–1	9–4	4–11	6–3	7–7	8–10

INDEX